Self-Sensing Concrete in Smart Structures

Self-Sensing Concrete in Smart Structures

Baoguo Han
Professor, School of Civil Engineering
Dalian University of Technology
Dalian, China

Xun Yu
Associate Professor, Department of Mechanical
and Energy Engineering
University of North Texas
Denton, TX, USA

Jinping Ou
Professor, School of Civil Engineering
Harbin Institute of Technology
Harbin, China
Professor, School of Civil Engineering
Dalian University of Technology
Dalian, China

AMSTERDAM • BOSTON • HEIDELBERG • LONDON
NEW YORK • OXFORD • PARIS • SAN DIEGO
SAN FRANCISCO • SINGAPORE • SYDNEY • TOKYO
Butterworth-Heinemann is an imprint of Elsevier

ELSEVIER

durability, limited sensing volume, and degradation of the structural performance of the concrete in case of embedded sensors.

The research of self-sensing concrete started in the early 1990s with the findings of sensing property of cement-based composites with short carbon fibers. Since then, much research work has been done on developing new types of self-sensing concrete with other functional fillers, such as ozone-treated carbon fiber, carbon-coated nylon fiber, metal-coated carbon fiber, steel fiber, graphite powder, nano TiO_2, Fe_2O_3, steel slag, carbon black, hybrid steel fiber and graphite powder, carbon nanotube, and hybrid carbon fiber and carbon black. Especially in recent years, a lot of attention has been paid to investigation on performance and structural application of self-sensing concrete prepared by adding such functional fillers as nickel powder, magnetic fly ash, hybrid magnetic fly ash and steel slag, carbon nanofiber, hybrid carbon fiber and carbon nanotube, hybrid carbon fiber and carbon black, hybrid carbon fiber and graphite powder, hybrid copper-coated carbon fiber and steel fiber, and hybrid iron containing conductive functional aggregate and carbon fiber.

Self-sensing concrete not only has potential in the field of structural health monitoring and condition evaluation for concrete structures, but also can be used for traffic detection, corrosion monitoring of rebar, military and border security, structural vibration control, and so on. It can ensure the safety, durability, serviceability, and sustainability of civil infrastructures such as high-rise buildings, large-span bridges, tunnel, high-speed railways, offshore structures, dams, and nuclear power plants.

In the past two decades, much effort has been made towards the advancement of self-sensing concrete, and many innovative achievements have been gained in both development and application of self-sensing concrete. This book includes three parts. The first part provides a systematical discussion on the structures of self-sensing concrete (Chapter 1), compositions of self-sensing concrete (Chapter 2), processing of self-sensing concrete (Chapter 3), fundamental sensing mechanism, measurement, and sensing properties of self-sensing concrete (Chapters 4–6), and structural application of self-sensing concrete (Chapter 7). The second part of the book presents the authors' research results in this area involving self-sensing concrete with carbon fiber, nickel powder, and carbon nanotube (Chapters 8–10). Finally, the third part discusses the future challenges for the development and deployment of self-sensing concrete and structures (Chapter 11).

As much of this book is based on the authors' previous researches, the authors want to thank the research team members in their groups. The authors also thank the funding supports from the National Science Foundation of China (51178148, 50808055, 50538020, 50420120133), the Ministry of Science and Technology of China (2011BAK02B01), Program for New Century Excellent Talents in University of China (NCET-11-0798), the USA National Science Foundation (CMMI-0856477), Federal Highway Administration (FHWA) of USA Department of Transportation (DTFH61-10-C-00011). The authors of this book also want to thank their families for their great support during the writing of this book.

Structures of Self-Sensing Concrete

Chapter Outline

1.1 Introduction and Synopsis

Self-sensing concrete (also called self-monitoring concrete, intrinsically smart concrete, and piezoresistive or pressure-sensitive concrete) is fabricated by adding functional fillers (carbon fibers, steel fibers, carbon nanotubes, nickel powder, etc.) into conventional concrete to increase its ability to sense strain, stress, cracking, or damage in itself while maintaining or even improving mechanical properties. Conventional concrete includes concrete (containing coarse and fine aggregates), mortar

1

(containing fine aggregates), and binder only (containing no aggregate, whether coarse or fine) in a generalized concept. It serves as a structural material with no or poor sensing ability. The presence of functional fillers enables the self-sensing property. The functional fillers need to be well-dispersed in a concrete matrix to form an extensive conductive network inside concrete. As the concrete material is deformed or stressed, the conductive network inside the material is changed, which affects the electrical parameters (e.g., electrical resistance, capacitance, and impedance) of the material. Strain (or deformation), stress (or external force), cracking, and damage under static and dynamic conditions can therefore be detected through measurement of the electrical parameters [1–5].

Structure–property relationships are at the heart of materials science. Self-sensing concrete, which has a highly complex structure, is a multiphase and multi-scale composite. Its structure covers over 10 orders of magnitude in size, ranging from nanometers (e.g., hydration product and some functional fillers) to micrometers (e.g., binder and some functional fillers), and then from millimeters (e.g., mortar and concrete) to tens of meters (final structures) [6–8]. Chapter 1 will introduce the structures of self-sensing concrete at different scale levels and their effects on the sensing properties of the composite.

1.2 Structures of Self-Sensing Concrete at the Macroscopic Level

At the macroscopic level, self-sensing concrete may be considered a two-phase material consisting of functional fillers dispersed in a concrete matrix, as shown in Figure 1.1.

In general, the functional filler phase usually exists in one of the three forms: fiber, particle, or a hybrid of fiber and particle. These fillers distribute in the concrete matrix phase to form a conductive network. As shown in Figure 1.2, fillers can be a variety of materials such as carbon fiber, carbon nanotube, steel fiber, nickel powder, graphite, or a hybrid of them. The concrete matrix phase, composed

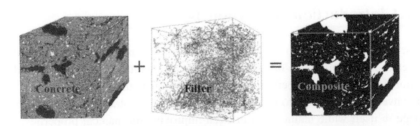

Figure 1.1 Structure of self-sensing concrete.

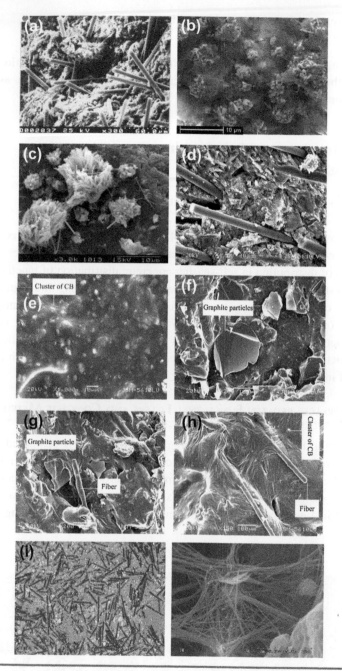

Figure 1.2 Scanning electron microscopy (SEM) photos of typical self-sensing concrete, (a) Cement concrete with carbon fiber [9], (b) Cement concrete with nickel powder [10], (c) Cement concrete with carbon nanotube [11], (d) Cement concrete with hybrid carbon fiber and graphite powder [12], (e) Asphalt concrete with carbon black [13], (f) Asphalt concrete with graphite [13], (g) Asphalt concrete with hybrid carbon fiber and graphite [13], (h) Asphalt concrete with hybrid carbon fiber and carbon black [13], (i) SEM images of carbon fiber and carbon nanotube in a cementitious matrix: (left) carbon fiber in cement composite (100 × magnification), (right) carbon nanotube bridging hydration products (5000 × magnification) [14].

of mineral aggregates glued together with a binder, supports the functional fillers and holds them in place. Here, the binder of concrete can be cement, asphalt, or even polymer [15,16].

1.3 Structures of Self-Sensing Concrete at the Microscopic Level

1.3.1 DISTRIBUTION OF FUNCTIONAL FILLERS IN CONCRETE MATRIX

At the microscopic level, the two phases of the structure of self-sensing concrete are not homogeneously distributed with respect to each other or to themselves. There are three levels of distribution in self-sensing concrete: distribution of functional fillers in binder, distribution of the binder with functional fillers among fine aggregates, and distribution of the fine aggregates with binder and functional fillers among coarse aggregates (as shown in Figure 1.3) [17]. Distribution of functional fillers in a concrete matrix is highly concerned with factors such as functional filler concentration, functional filler geometrical shape, and processing methods, which will be introduced in detail in Chapters 3 and 5.

1.3.2 INTERFACES BETWEEN FUNCTIONAL FILLERS AND CONCRETE MATRIX

There is also a third phase in self-sensing concrete, which is composed of the interfaces between functional fillers and concrete matrix and those between functional fillers [6]. Because functional fillers are mainly micro-scale or nano-scale, the potential filler–matrix and filler–filler interface areas are enormous. These interfaces affect electrical contact between fillers and concrete matrix and among fillers (as

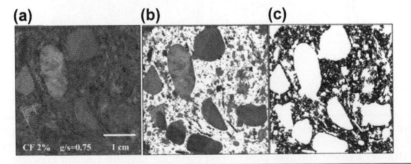

Figure 1.3 Photos of concrete with carbon fibers: (a) image without digital processing; (b) image with digital processing to highlight aggregates as dark regions; (c) image with digital processing to highlight cement paste as dark regions [17].

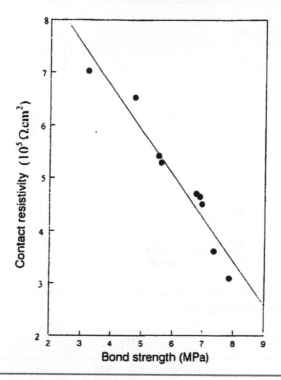

Figure 1.4 Variations in contact electrical resistivity with bond strength at 28 days of curing [18].

shown in Figure 1.4) [18], thereby affecting the conductive network and electrical conductivity of self-sensing concrete. Therefore, they will have a great influence on the sensing behavior of self-sensing concrete.

For example, Fu and Chung observed that the self-sensing behavior of carbon fiber cement mortar at a curing age of 7 days is entirely different from that at a curing age of 14 days and 28 days (as shown in Figure 1.5 [19]). They considered this phenomenon to result from weakening of the fiber–cement interface as curing progresses [20].

Fu et al. enhanced the interfacial bond between fiber and matrix by ozone treatment of the fibers, thus improving the strain-sensing ability of carbon fiber–reinforced cement (as shown in Figure 1.6). The improvement pertains to better repeatability upon repeated loading and an increased strain sensitivity coefficient [21].

Li et al. stated that the surface of carbon nanotube treated with a mixed solution of H_2SO_4 and HNO_3 is covered by C-S-H. As a result, there are many fewer contact points of treated carbon nanotube in composites than those of untreated carbon nanotube in cement composites, which contributes to the higher

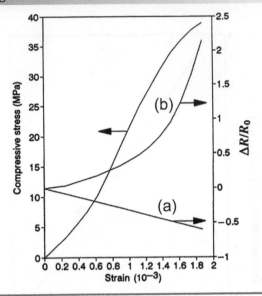

Figure 1.5 Sensing properties of self-sensing concrete with carbon fibers (a) 28 days of curing; (b) 7 days of curing [19].

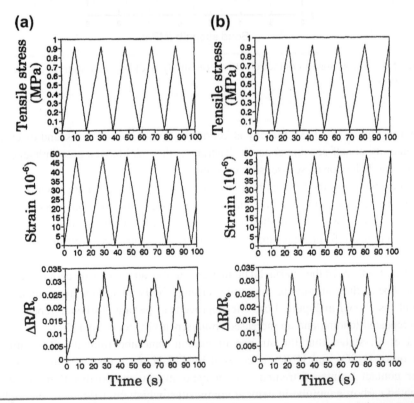

Figure 1.6 Comparison of sensing properties of self-sensing concrete with as-received carbon fibers and ozone-treated carbon fibers [21], (a) with as-received carbon fibers, (b) with ozone-treated carbon fibers.

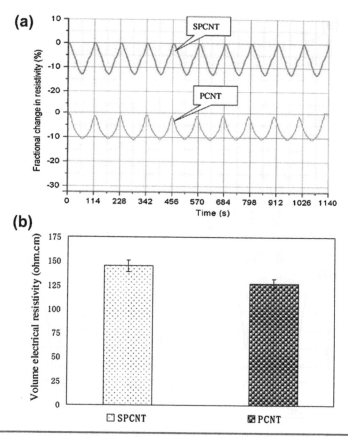

Figure 1.7   Comparison of sensing and conductive properties of self-sensing concrete with as-received carbon nanotubes and surface treatment carbon nanotubes (SPCNT are surface treatment carbon nanotubes, and PCNT are as-received carbon nanotubes) [11], (a) Sensing property, (b) Conductive property.

compressive sensitive properties and lower electrical conductivity (as shown in Figure 1.7) [11].

Yu and Kwon observed that the sensing property of cement composites with carbon nanotubes treated with a mixed solution of H_2SO_4 and HNO_3 has a higher signal-to-noise ratio compared with that of carbon nanotube cement composites fabricated with surfactant (as shown in Figure 1.8 and Table 1.1). They pointed out that the difference in sensing properties between two composites can be attributed to the different nanotube-to-nanotube interfaces. Carbon nanotubes treated with a mixed solution of H_2SO_4 and HNO_3 can contact directly with each other in the carbon nanotube network. However, if carbon nanotube surfaces are wrapped with surfactants, contact between carbon nanotubes can be blocked by the surfactant, which results in the lower signal-to-noise ratio [22].

functional fillers and concrete matrix in self-sensing concrete. The effect of water on the sensing property of self-sensing concrete will be described in detail in Chapters 5 and 6 [24–32].

▌ 1.4 Summary and Conclusions

Self-sensing concrete is a composite material whose microstructure contains random features over a wide range of length scales, from nanometers to several meters, with each length scale presenting a new random composite. Self-sensing concrete consists of a concrete matrix phase, a functional filler phase, and an interface phase between fillers and matrix. Its structure is highly heterogeneous, complex, and dynamic. The sensing properties of self-sensing concrete are closely related to its structure, especially the distribution of functional fillers in a concrete matrix, the interfaces between functional fillers and the cement matrix, and void and liquid phases in a concrete matrix.

▌ References

[1] Han BG, Yu X, Ou JP. Chapter 1: multifunctional and smart carbon nanotube reinforced cement-based materials. In: Gopalakrishnan K, Birgisson B, Taylor P, Attoh-Okine NO, editors. Book: nanotechnology in civil infrastructure: a paradigm shift, vol.1–47. Publisher: Springer; 2011. p. 276.

[2] Chung DDL. Self-monitoring structural materials. Mater Sci Eng Rep 1998;22(2):57–78.

[3] Hou TC, Lynch JP. Conductivity-based strain monitoring and damage characterization of fiber reinforced cementitious structural components. Proc SPIE 2005;5765:419–29.

[4] Mao QZ, Zhao BY, Sheng DR, Li ZQ. Resistance changement of compression sensible cement speciment under different stresses. J Wuhan Univ Technol 1996;11:41–5.

[5] Han BG. Properties, sensors and structures of pressure-sensitive carbon fiber cement paste. Dissertation for the Doctor Degree in Engineering, Harbin Institute of Technology; 2006.

[6] Mehta PK, Monteiro PJM. Concrete: microstructure, properties and materials. New York: McGraw-Hill; 2006.

[7] Shetty MS. Concrete technology: theory and practice. S. Chand and Company, Limited; 2000.

[8] Boyd AJ, Mindess S. Cement and concrete: trends and challenges. American Ceramics Society; 2002.

[9] Wang XF, Wang YL, Jin ZH. Electrical conductivity characterization and variation of carbon fiber reinforced cement composite. J Mater Sci 2002;37:223–7.

[10] Han BG, Han BZ, Yu X. Experimental study on the contribution of the quantum tunneling effect to the improvement of the conductivity and piezoresistivity of a nickel powder-filled cement-based composite. Smart Mater Struct 2009;18:065007 (7pp).

[11] Li GY, Wang PM, Zhao XH. Pressure-sensitive properties and microstructure of carbon nanotube reinforced cement composites. Cem Concr Compos 2007;29:377–82.

[12] Fan XM, Fang D, Sun MQ, Li ZQ. Piezoresistivity of carbon fiber graphite cement-based composites with CCCW. J Wuhan Univ Technol-Materials Sci Ed 2011;25(2):339–43.

[13] Wu SP, Mo LT, Shui ZH, Chen Z. Investigation of the conductivity of asphalt concrete containing conductive fillers. Carbon 2005;43:1358–63.

[14] Azhari F, Banthia N. Cement-based sensors with carbon fibers and carbon nanotubes for piezoresistive sensing. Cem Concr Compos 2012;34:866–73.

[15] Sett K. Characterization and modeling of structural and self-monitoring behavior of fiber reinforced polymer concrete. Dissertation for the Master of Science in Civil Engineering, University of Houston, USA; 2003.

[16] Prashanth P, Vipulanandan C. Characterization of thin disk piezoresistive smart material for hurricane applications. THC-IT 2009 conference and exhibition; 2009. 1–2.

[17] Baeza FJ, Chung DDL, Zornoza E, Andión LG, Garcés PG. Triple percolation in concrete reinforced with carbon fiber. ACI Material J 2010;107(4):396–402.

[18] Fu XL, Chung DDL. Contact electrical resistivity between cement and carbon fiber: its decrease with increasing bond strength and its increase during fiber pull-out. Cem Concr Res 1995;25(7):1391–6.

[19] Fu XL, Chung DDL. Effect of curing age on the self-monitoring behavior of carbon fiber reinforced mortar. Cem Concr Res 1997;27(9):1313–8.

[20] Chung DDL. Piezoresistive cement-based materials for strain sensing. J Intell Mater Syst Struct 2002;13(9):599–609.

[21] Fu XL, Lu W, Chung DDL. Improving the strain sensing ability of carbon fiber reinforced cement by ozone treatment of the fibers. Cem Concr Res 1998;28(2):183–7.

[22] Yu X, Kwon E. Carbon-nanotube/cement composite with piezoresistive property. Smart Mater Struct 2009;18:055010 (5pp).

[23] Azhari F. Cement-based sensors for structural health monitoring. Dissertation for the Master Degree of Applied Science. Canada: University of British Columbia; 2008.

[24] Jia XW. Electrical conductivity and smart properties of Fe1-σO waste mortar. Dissertation for the Doctor Degree in Engineering. China: Chongqing University; 2009.

[25] Li CT. Study on conductivity and strain sensitivity of steel-slag concrete. Dissertation for the Master Degree in Engineering. China: Chongqing University; 2004.

[26] Wang YL, Zhao XH. Positive and negative pressure sensitivities of carbon fiber-reinforced cement-matrix composites and their mechanism. Acta Mater Compos Sin 2005;22(4):40–6.

[27] Han BG, Zhang LY, Ou JP. Influence of water content on conductivity and piezoresistivity of cement-based material with both carbon fiber and carbon black. J Wuhan Univ Technol-Mater Sci Ed 2010;25(1):147–51.

[28] Han BG, Yu X, Ou JP. Effect of water content on the piezoresistivity of CNTs/cement composites. J Mater Sci 2010;45:3714–9.

[29] Tashiro C, Ishida H, Shimamura S. Dependence of the electrical resistivity on evaporable water content in hardened cement pastes. J Mater Sci Lett 1987;6:1379–81.

[30] Zhang ZG. Functional composite materials. Beijing: Chemical Industry Press; 2004.

[31] Tang DS, Ci LJ, Zhou WZ, Xie SS. Effect of H_2O adsorption on the electrical transport properties of double-walled carbon nanotubes. Carbon 2006;44:2155–9.

[32] Na PS, Kim H, So HM, Kong KJ, Chang H, Ryu BH, et al. Investigation of the humidity effect on the electrical properties of single-walled carbon nanotube transistors. Appl Phys Lett 2005;87:093101 (3pp).

2.1 Introduction and Synopsis

The structure of self-sensing concrete depends to a high degree on the composition of the composites. As a composite, self-sensing concrete consists mostly of matrix materials (i.e., conventional concrete materials) and functional filler. In addition, some auxiliary materials may be necessary to disperse functional fillers into matrix materials. The available composition materials for fabricating self-sensing concrete are varied. Therefore, selection of suitable materials and determination of their proportions are important for fabricating self-sensing concrete. Chapter 2 will introduce the composition of self-sensing concrete (including matrix material, functional filler, materials to aid filler dispersion, and the mixing proportion design), and the relationships between composition materials and the properties of the composites.

2.2 Matrix Material

The matrix material is the component that holds the functional filler together to form the bulk of the composite, so all types of concrete can be used as a matrix for self-sensing concrete. Here, concrete is a generalized concept that includes concrete (containing coarse and fine aggregates), mortar (containing fine aggregates), and paste (containing no aggregate, whether coarse or fine). In previous studies, typical Portland cement concrete (including cement concrete, cement mortar, and cement paste) was most frequently used as the matrix material for self-sensing concrete because Portland cement is the most widely used binder material. Recently, some new types of Portland cement concrete have been chosen as a matrix to develop self-sensing concrete. For example, Hong employed slurry-infiltrated fiber concrete as the matrix to obtain self-sensing concrete with high mechanical properties [1]. Fan used cementitious capillary crystalline waterproofing material as the matrix of self-sensing concrete to combine self-healing waterproofing with self-sensing ability [2]. Hou et al. and Lin et al. incorporated conductive fillers including carbon fiber, steel fiber, and carbon black into the engineered cementitious composite to enhance its self-sensing behavior while maintaining its tensile strain-hardening behavior [3,4]. Besides Portland cement, Cheng et al. tried to adopt sulphoaluminate cement as the binder in making self-sensing concrete [5]. Saafi et al. used geopolymer cement as a binder to fabricate self-sensing concrete with carbon nanotubes [6].

Since the potential application of self-sensing concrete in traffic detection was recognized, the use of asphalt concrete as a matrix to develop self-sensing concrete has increasingly been paid attention. Comprehensive research into intrinsic self-sensing asphalt concrete has been successfully performed [7–15]. In addition, Sett employed polymer concrete as a matrix to make self-sensing concrete [16]. Because previous research focused mainly on the sensing property of self-sensing concrete fabricated with a cement binder, in this chapter, self-sensing concrete refers to cement concrete unless otherwise specified.

Although the concrete matrix has no or poor sensing ability and contributes only slightly to the electrical conduction of the whole composite system, some properties of the concrete matrix have substantial effects on the sensing properties of self-sensing concrete. This is because the sensing ability of self-sensing concrete is strongly related to its mechanical behavior (i.e., stress, strain, damage, etc.) and electrical conduction, whereas the mechanical properties (e.g., ultimate stress and strain, Young's modulus, and Poisson ratio) of self-sensing concrete depend heavily on the mechanical properties of the concrete matrix. In addition, the type and mixing proportions of the materials chosen as the matrix to prepare concrete also influence the dispersion of functional fillers, the distribution of functional fillers in the matrix, and the mechanical properties of the composites, thus affecting the sensing abilities of the composites. For example, Mao et al. observed that the sensing property of concrete with carbon fiber deteriorates with an increase in the water–cement ratio, and moderate heat dam cement paste with carbon fiber has better sensing properties compared with slag cement with carbon fiber [17]. Chen and Chung stated that the linearity of sensing properties is better for cement mortars containing methylcellulose than those containing hybrid methylcellulose and silica fume or latex [18]. Li [19] and Han et al. [20] suggested that increasing the water cement ratio can improve the self-sensing sensitivity of steel–slag concrete (as shown in Figure 2.1 [21]) and carbon nanotube–cement composites. Two factors contribute to the effect of the water–cement ratio on the sensing sensitivity of self-sensing concrete. One is the deformation capacity of the concrete matrix;

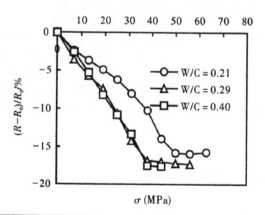

Figure 2.1 Sensing properties of self-sensing concrete containing steel–slag with different water–cement ratios (W/C) [21].

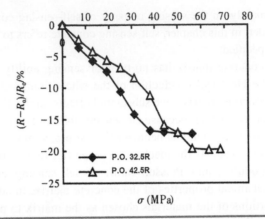

Figure 2.2 Sensing properties of self-sensing concrete containing steel–slag fabricated with different cement strength grades [21].

the other is the dispersion of fillers in the concrete matrix. Concrete fabricated with a higher water–cement ratio has larger deformation than that fabricated with a lower water–cement ratio at the same compressive stress, so it is easier to change the conductive network in the former. This indicates that the electrical resistance of the composite with a higher water–cement ratio is more sensitive to loading. In addition, a higher water–cement ratio is beneficial for the dispersion of fillers in a concrete matrix [19,20]. Li and Jia observed that an increase in cement strength grade (as shown in Figure 2.2) [21] and the use of a water-reducing agent [19,22] cause a decrease in sensitivity of the steel–slag concrete, because they improve the strength of the concrete matrix and decrease its deformation capacity.

In addition, typical aggregates are electrically insulating in nature, and they will create some obstacles or cut the electron flow through the conductive network. The adoption of aggregates will decrease the conductivity of self-sensing concrete (as shown in Figure 2.3) [23,24]. Therefore, the concentration level of functional filler must be increased to achieve the expected self-sensing ability when aggregates, especially coarse ones, are included in the composites. All of these examples illustrate that the concrete matrix would affect the deformation capacity and inside conductive network of the composite, thus affecting the sensing property of self-sensing concrete.

2.3 Functional Filler

Functional filler is an essential and critical component of self-sensing concrete, because it dominates the sensing property of the self-sensing concrete. Since the sensing behavior of the self-sensing concrete was first observed, researchers have

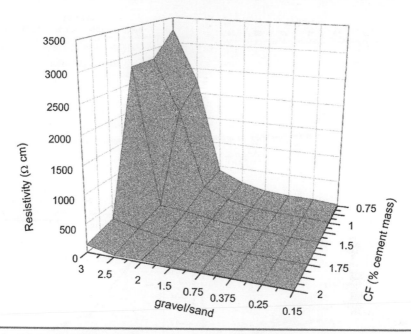

Figure 2.3   Dependence of electrical resistivity of concrete on carbon fiber proportion and gravel–sand ratio [24].

been trying to find or develop new functional fillers that can endow conventional concrete with high self-sensing performance.

2.3.1 Types of Functional Filler

By now, more than 10 types of functional fillers have proved effective for enhancing the sensing performance of concrete. In addition, a hybrid of two or several types of functional fillers can bring out preferable sensing properties of self-sensing concrete, which cannot be achieved by any of the functional fillers alone. This can contribute to the complementary effect of different functional fillers on the modification of sensing properties. For example, Azhari and Banthia observed that concrete with hybrid carbon fiber and carbon nanotube can provide better signal quality, improve reliability, and increase sensitivity over concrete carrying carbon fiber alone (as shown in Figure 2.4). The superior sensing performance of concrete with hybrid fillers may be because carbon nanotubes help close the gaps in the carbon fiber conduction path and provide the concrete with a more homogeneous quality in terms of electrical properties [25].

Jia stated that concrete with hybrid magnetic fly and steel slag has a higher sensitivity than that with magnetic fly or steel slag only [22]. Ou and Han observed

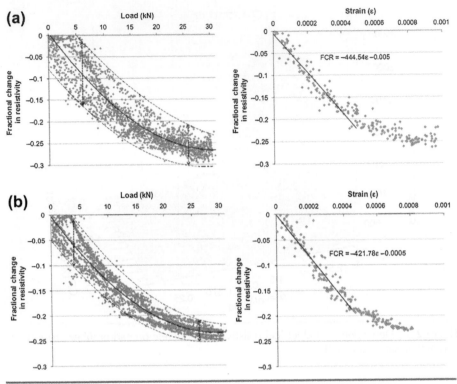

Figure 2.4 Comparison of sensing properties of self-sensing concrete with carbon fiber only and with hybrid carbon fiber and carbon nanotube [25], (a) with carbon fiber, (b) with hybrid carbon fiber and carbon nanotube.

that a hybrid of carbon fiber and carbon black enhances the reproducibility and linearity of the sensing performance. They attributed this phenomenon to the combination of fibrous and particle fillers, which causes the transportation of charges over long and short distances and enhances the contacting and tunneling conduction effect [26]. Luo et al. observed that a hybrid of carbon nanotube and carbon fiber is more effective in improving the repeatability and stability of self-sensing properties of the composites rather than the sensitivity [27]. However, the hybrid of carbon nanotube and carbon black was found to be effective in enhancing the stress–strain sensitivity of the composites, but almost no help for improving the repeatability and stability of self-sensing properties of the composites [28]. Fan et al. found that a hybrid of carbon fiber and graphite powder can give stable electrical conductivity and sensing sensitivity [29]. Deng stated that concrete with a hybrid of iron-containing conductive functional aggregate and carbon fiber has a higher sensing sensitivity than that with carbon fiber only [30]. Lin et al. observed that the incorporation of carbon black is capable of decreasing the initial electrical resistivity of

TABLE 2.1 Improvement of Hybrid Functional Fillers to Sensing Properties of Self-Sensing Concrete

Hybrid Fillers	Improved Parameters	Compared Filler
Carbon fiber and carbon nanotube [25]	Sensing reliability and sensitivity	Carbon fiber alone
Magnetic fly ash and steel slag [21]	Sensing sensitivity	Magnetic fly ash or steel slag alone
Carbon fiber and carbon black [26]	Sensing reproducibility and linearity	Carbon fiber alone
Carbon fiber and carbon nanotube [27]	Sensing repeatability and stability	Carbon fiber alone
Carbon nanotube and carbon black [28]	Sensing sensitivity	Carbon nanotube alone
Carbon fiber and graphite powder [29]	Stability of conductivity and sensing sensitivity	Carbon fiber alone
Iron-containing conductive functional aggregate and carbon fiber [30]	Sensing sensitivity	Carbon fiber alone
Polyvinyl alcohol fiber and carbon black [31]	Background resistivity and sensing sensitivity	Polyvinyl alcohol fiber alone

engineered cementitious composite with polyvinyl alcohol fiber, which is essential for lowering the background resistivity and enhancing the resistivity change in the composites [31]. The improvement of hybrid functional fillers to the sensing properties of self-sensing concrete is summarized in Table 2.1.

All of these functional fillers can be classified into different categories according to different criteria. Table 2.2 summarizes the classification of these functional fillers.

The self-sensing behavior of concrete depends heavily on several parameters of the functional fillers, such as material components, morphology (e.g., shape, size, length, surface state, degree of aggregation and agglomeration), and concentration level. Research results for self-sensing concrete show that fibrous fillers with a high aspect ratio (i.e., ratio of length to diameter) can modify the sensing ability of concrete at a lower concentration compared with particle fillers (e.g., steel slag, as shown in Figure 2.5 [32]). The effective concentration is not higher than 1.5% for fibrous fillers, whereas it is at least 5% for particle fillers. In addition, fibrous fillers can improve the ductility and toughness of cement-based materials without sacrificing other properties. The presence of fibrous fillers also controls cracking of concrete, so that the cracks do not begin and propagate catastrophically, as in the case

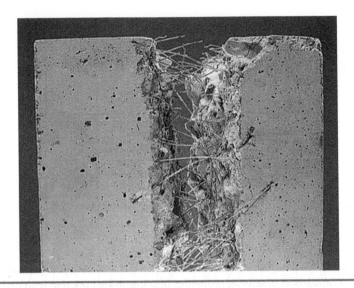

Figure 2.6 Concrete with steel fibers [34].

(as shown in Figure 2.6 [34]). Basically, steel fiber can be categorized into five groups, depending on the manufacturing process and its shape and/or section: cold-drawn wire, cut sheet, melt-extracted, mill cut, and modified cold-drawn wire (as shown in Figure 2.7 [35]). In 2003, Wen and Chung first fabricated cement paste with self-sensing properties using steel fibers with a length of 6 mm and diameter of 8 μm (as shown in Figure 2.8 and Figure 2.9 [36]) [37]. Hong employed steel fibers with a length of 32 mm and diameter of 0.64 mm to develop self-sensing concrete [1]. Hou and Lynch also developed an engineered cementitious composite with sensing properties by incorporating steel fibers [3]. Teomete and Kocyigit used steel fiber with a length of 6 mm to fabricate self-sensing concrete with tensile strain-sensing properties [38].

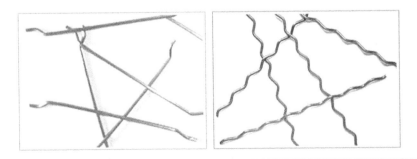

Figure 2.7 Examples of steel fiber [35].

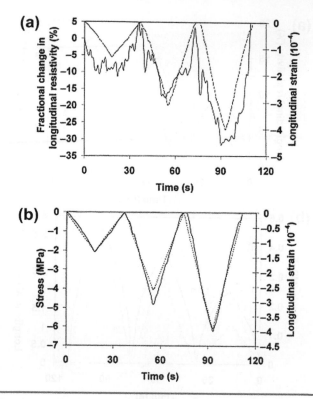

Figure 2.8 Sensing properties of concrete with 0.36 vol. % steel fiber under compression: variations in fractional change in electrical resistivity (solid curve) with strain (dashed curve) (a), and in the strain (solid curve) with stress (dashed curve) (b) [36].

2.3.2.2 Carbon Black

Carbon black is a form of amorphous carbon that has a high surface-area-to-volume ratio. Carbon black is categorized as acetylene black, channel black, furnace black, lampblack, or thermal black, according to its manufacture process. The type of carbon black can also be characterized by the size distribution of the primary particles, the degree of their aggregation and agglomeration (as shown in Figure 2.10 [39]), and the chemicals adsorbed onto the particle surfaces. The average primary particle diameters of several commercially produced carbon blacks range from 10 to 400 nm, whereas the average diameters of carbon black aggregates range from 100 to 800 nm [40]. As a kind of carbon material, carbon black has advantages such as light weight, high chemical and thermal stability, permanent electrical conductivity, and low cost. In 2006, Li et al. first employed 120-nm furnace black to develop carbon black cement paste with strain-sensing properties (as shown in Figure 2.11(a))

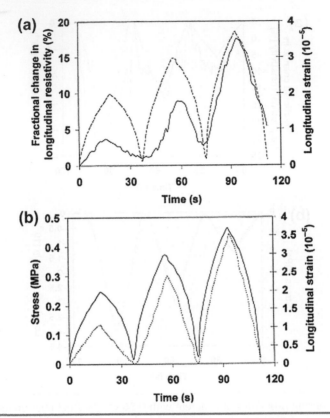

Figure 2.9 Sensing properties of concrete with 0.36 vol. % steel fiber under tension: variations in the fractional change in electrical resistivity (solid curve) with strain (dashed curve) (a), and in the strain (solid curve) with stress (dashed curve) (b) [36].

[41,42]. In 2008, Han et al. obtained a carbon black cement paste with piezoresistivity (as shown in Figure 2.11(b)) by adding acetylene black into cement paste [43]. Long, Gongshu, Lin et al., and Zhao et al. fabricated self-sensing concrete by incorporating carbon black about 30 nm in diameter into conventional concrete (as shown in Figure 2.12) [4,44–46]. The concentration of carbon black used in their study was below 1%, much lower than that used in Refs [41] and [43] and similar to fibrous fillers. Nowadays, carbon black is one of the most commonly used functional fillers to incorporate sensing capabilities into concrete.

2.3.2.3 Steel Slag

Steel slag is an industrial byproduct obtained from the steel manufacturing industry. It is produced in large quantities during steel-making operations that use electric arc furnaces. Steel slag can also be produced by smelting iron ore in a basic oxygen

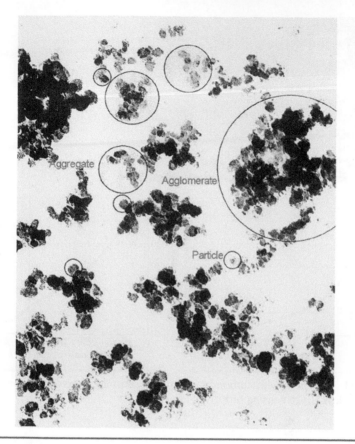

Figure 2.10 Carbon black [39].

furnace. According to methods for cooling molten steel slag, steel slag is classified into five types: natural air-cooling steel slag, water-spray steel slag, water-quenching steel slag, air-quenching steel slag, and shallow box chilling steel slag [47,48]. Most steel slag contains a high content of $Fe_{1-\sigma}O$ and other metal oxides. $Fe_{1-\sigma}O$ includes FeO, Fe_2O_3, and Fe_3O_4, all of which are nonstoichiometric compounds, so $Fe_{1-\sigma}O$ has the properties of a semiconductor. The electrical resistivity of FeO and Fe_3O_4 is 5×10^{-2} and $4 \times 10^{-3}\,\Omega\,cm$, respectively, which is basically the same as that of pitch-based carbon fiber. As a result, steel slag presents good electrically conductive properties [23]. In addition, steel slag can be used as aggregates in concrete to replace natural aggregates, because it has favorable mechanical properties, including strong bearing and shear strength, good soundness characteristics, and high resistance to abrasion and impact. Steel slag aggregates are fairly angular, roughly cubical pieces with a flat or elongated shape (as shown in Figure 2.13 [49]). They have a rough vesicular nature with many non-interconnected cells, which gives a

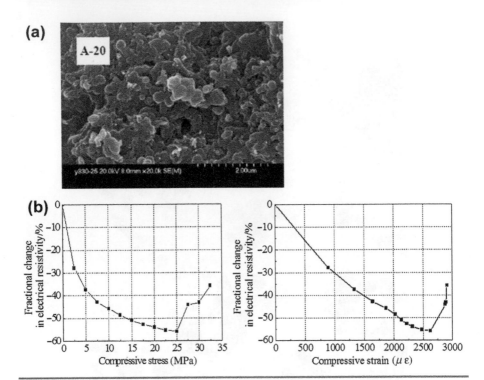

Figure 2.11 Concrete with carbon black and its sensing property, (a) Concrete with carbon black [41], (b) Sensing property [43].

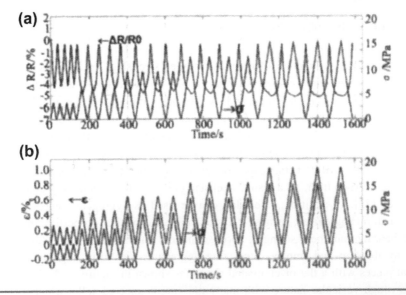

Figure 2.12 Sensing behavior of concrete with 0.5% carbon black, (a) Fractional change in resistance ($\triangle R/R$) and variation of stress (σ) with time, (b) Variation of stress and strain (ε) with time [46].

Figure 2.13 Steel slag [49].

greater surface area than smoother aggregates of equal volume. This feature provides an excellent bond with concrete binder. Replacing some or all natural aggregates with steel slag is helpful for reducing environmental pollution and the consumption of resources [47,48]. Therefore, steel slag is a promising kind of filler because it works as both functional filler and aggregate. The incorporation of air-quenching steel slag of 0.315–5 mm (in which the content of $Fe_{1-\sigma}O$ is over 30%) into concrete to fabricate mechanically sensitive concrete was investigated by Li et al. in 2005 [21]; subsequently, Jia performed a systematic study of this concrete (as shown in Figure 2.14) [23,50].

2.3.2.4 Graphite Powder

Graphite is a polymorph carbon. Graphite has a layered, planar structure. In each layer, the carbon atoms are arranged in a honeycomb lattice with separation of 0.142 nm; the distance between planes is 0.335 nm. Graphite is soft because the covalent bonds of the atoms are not tightly packed together. Depending on the mode of occurrence and origin, it is graded into three forms: flake, crystalline (lumpy), and cryptocrystalline (amorphous) (as shown in Figure 2.15) [51,52]. Graphite is a good conductor of heat and electricity. It has high refractoriness and stable chemical properties. Liu and Wu adopted graphite powder to fabricate asphalt concrete with sensing properties (as shown in Figure 2.16) [53]. Hong employed graphite powder to develop self-sensing slurry–infiltrated steel fiber concrete [1].

2.3.2.5 Carbon Nanofiber

Carbon nanofiber is a kind of nanoscale carbon fiber. It is a quasi-one-dimensional carbon material between carbon nanotube and carbon fiber (as shown in Figure 2.17)

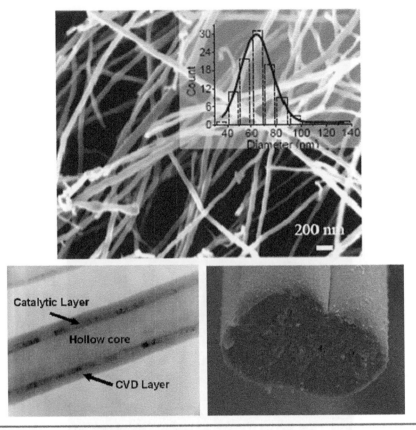

Figure 2.17 Images of carbon nanofibers [54].

2.4 Dispersion Material

The functional fillers used to fabricate self-sensing concrete are usually on the micron or nanoscale, which makes their dispersion into matrix challenging. As the surface area of fillers increases, the attractive forces between the fillers increase [64]. Especially for fibrous fillers, high aspect ratios combined with high flexibilities increase the possibility of fiber entanglement and close packing [65]. Therefore, some effective dispersion materials are needed to help the dispersion of functional fillers in concrete matrix and improve the homogeneity of the concrete matrix. The usage of dispersion materials has three benefits: one may obtain reproducible and stable sensing and mechanical properties, achieve full realization of the improvement effects of functional fillers (i.e., decreasing the concentration level of fillers), and decrease the consumption of mechanical mixing energy. An ideal dispersion material should have good compatibility with the component

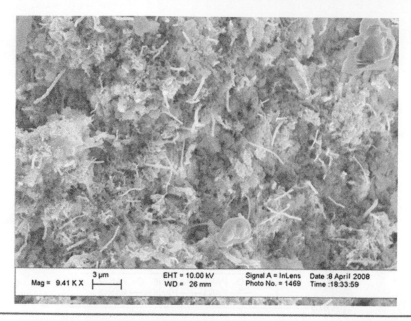

Figure 2.18   Scanning electron microscopy photo of concrete with carbon nanofiber [57].

materials of the matrix, i.e., no or little negative effect on cement hydration, workability, and mechanical properties of the composites [58]. Dispersion materials can be classified into two types: namely, surfactant and mineral admixture. The dispersion capability of surfactant is achieved by wetting, electrostatic repulsion, and/or steric hindrance effects, whereas for a mineral admixture it is achieved by gradation, adsorption, and/or separation effects [66].

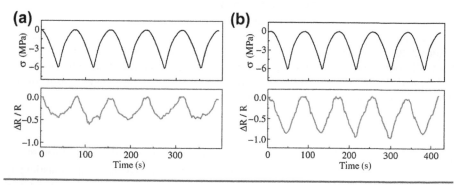

Figure 2.19   Sensing properties of concrete with carbon nanofiber: (a) with 0.5% carbon nanofiber with a length of 50–200 μm and diameter of 150 nm; (b) with 0.5% carbon nanofiber with a length of 50–200 μm and diameter of 100 nm [58].

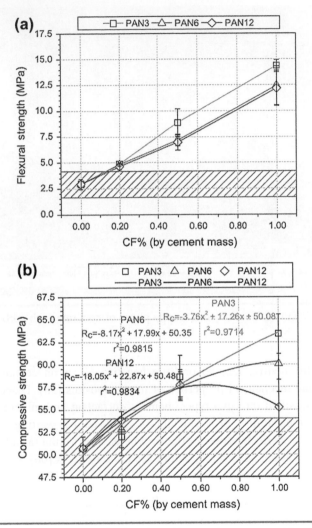

Figure 2.20 Mechanical strength of carbon fiber cement paste with different carbon fiber (CF) percentages (by mass of cement) and different carbon fiber lengths (PAN3, PAN6, and PAN12 denote carbon fiber of 3-, 6-, and 12.5-mm lengths, respectively) [60], (a) Flexural strength, (b) Compressive strength.

Some raw materials used to fabricate concrete matrix are also dispersion materials. They include water-reducing agents (e.g., polycarboxylate superplasticizer, naphthalene sulfonic acid, and formaldehyde condensate superplasticizer), silicon fume, and fly ash. These dispersion materials not only improve the dispersion of functional fillers, they are also beneficial for the dispersion of other components into

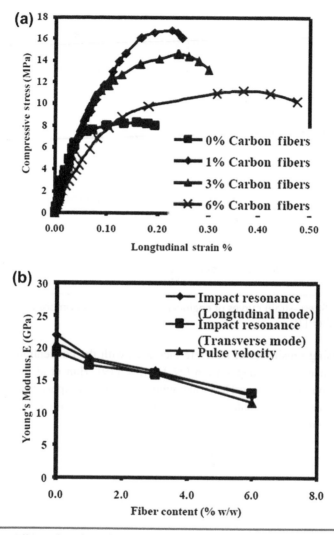

Figure 2.21 Effect of carbon fiber on the mechanical properties of concrete [61], (a) Stress–strain relationship, (b) Young's modulus.

the concrete matrix. Therefore, they possess double dispersion effects, which is beneficial for homogeneity and the sensing properties of self-sensing concrete. Wang et al. stated that the addition of silicon fume can improve the dispersion of carbon fiber in matrix [67]. Li observed that the use of silicon fume and fly ash can improve the self-sensing sensitivity of steel–slag concrete. Addition of 5% silicon fume greatly enhances the self-sensing sensitivity of the concrete, but a continuous increase of silicon fume has no more benefit when its concentration is more than 5%.

TABLE 2.4 Effective Dispersion Materials for Different Functional Fillers

Category	Dispersion Material	Dispersion Objects	Dispersed Functional Fillers
Surfactant	Water-reducing agent	Functional fillers and cement particles	Carbon nanotube, carbon nanofiber, carbon black, graphite powder, nickel powder, nano TiO_2, nano Fe_2O_3
	Methylcellulose	Functional fillers	Carbon fiber, carbon nanotube
	Carboxy methylcellulose	Functional fillers	Carbon fiber
	Carboxyethyl cellulose	Functional fillers	Carbon fiber
	Hydroxypropyl methylcellulose	Functional fillers	Carbon fiber
	Sodium dodecyl sulfate	Functional fillers	Carbon nanotube, carbon nanofiber
	Sodium dodecylbenzene sulfonate	Functional fillers	Carbon nanotube
	Sodium dodecylbenzene sulfonate and polyacrylic acid	Functional fillers	Carbon nanotube
Mineral admixture	Silicon fume	Functional fillers and cement particles	Carbon fiber, carbon black, steel fiber, nickel powder
	Fly ash	Functional fillers and cement particles	Steel slag, magnetic fly

properties are complex, some effective experiment design methods, such as orthogonal design and uniform design, are necessary to obtain a suitable or optimal mixing proportion design formula [70].

2.6 Summary and Conclusions

The compositions of self-sensing concrete are relatively complex compared with conventional concrete. The materials used to fabricate self-sensing concrete include matrix material, functional filler, and materials to aid filler dispersion. Matrix materials are the component that holds the functional filler and mainly contributes to mechanical properties. Generally, all type of concrete can be used as a matrix for self-sensing concrete. Functional filler is an essential component of self-sensing concrete. To date, more than 10 types of functional filler and hybrids of two or several types of functional fillers have proved effective in fabricating self-sensing concrete. Some effective dispersion materials are needed to help disperse

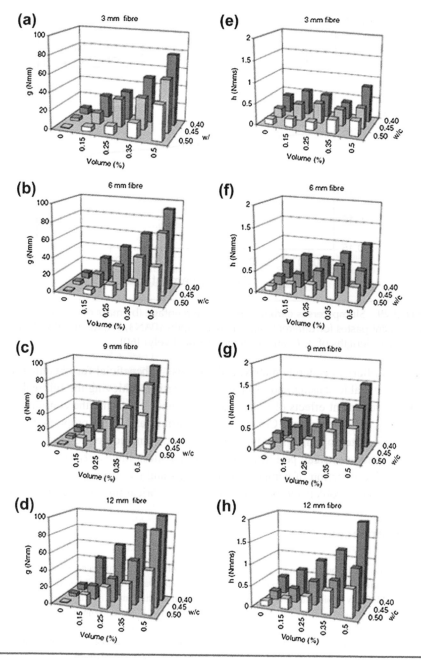

Figure 2.28 Effect of carbon fiber on yield stress and plastic viscosity of fresh cement mortar with carbon fiber (a-d): yield stress of the composite with 3 mm fiber, 6 mm fiber, 9 mm fiber, and 12 mm fiber, respectively, (e-h): plastic viscosity of the composite with 3 mm fiber, 6 mm fiber, 9 mm fiber, and 12 mm fiber, respectively. (Note: g [the intercept] is proportional to yield stress, and h [the gradient] is proportional to plastic viscosity of the material.) [69].

[30] Deng X. Preparation and performance investigation of iron containing aggregate and its cement-based conductive composites [dissertation for the master degree in engineering]. China: Wuhan University of Technology; 2011.

[31] Lin VWJ, Mo L, Lynch JP, Li VC. Mechanical and electrical characterization of self-sensing carbon black ECC. Proc SPIE – Int Soc Opt Eng 2011;7983:798316. 12 pp.

[32] Jia XW, Qian JS, Tang ZQ. Research and mechanism analysis on the compression sensitivity of steel slag concrete. Mater Sci Technol 2010;18(1):66–70.

[33] Banthia N. Fiber reinforced concrete for sustainable and intelligent infrastructure. First International Conference on Sustainable Built Environment Infrastructures in Developing Countries, Algeria; 2009. 337–350.

[34] http://www.civilengineeringgroup.com/steel-fiber-reinforced-concrete-2.html.

[35] http://www.steelfiber.org/.

[36] Chung DDL. Piezoresistive cement-based materials for strain sensing. J Intell Mater Syst Struct 2002;13(9):599–609.

[37] Wen SH, Chung DDL. A comparative study of steel fiber cement and carbon fiber cement as piezoresistive strain sensors. Adv Cem Res 2003;15(3):119–28.

[38] Teomete E, Kocyigit OI. Tensile strain sensitivity of steel fiber reinforced cement matrix composites tested by split tensile test. Constr Build Mater 2013;47:962–8.

[39] http://www.grin.com/en/doc/243611/new-application-of-crystalline-cellulose-in-rubber-composites.

[40] http://www.inchem.org/documents/iarc/vol65/carbon.html.

[41] Li H, Xiao HG, Ou JP. Effect of compressive strain on resistance of carbon black filled cement-based composites. Cem Concr Compos 2006;28:824–8.

[42] Xiao HG. Piezoresistivity of cement-based composites filled with nanophase materials and self-sensing smart structural system [dissertation for the doctor degree in engineering]. China: Harbin Institute of Technology; 2006.

[43] Han BG, Chen W, Ou JP. Study on piezoresistivity of cement-based materials with acetylene carbon black. Acta Mater Compos Sin 2008;25(3):39–44.

[44] Long X. Research on pressure-sensitivity of compound material of carbon black filled cement [dissertation for the master degree in engineering]. China: Wuhan University of Technology; 2007.

[45] Gongshu Q. Study on long-term mechanical properties and pressure-sensitivity of concrete containing nano-sized carbon black or carbon fibers [dissertation for the master degree in engineering]. China: Shantou University; 2009.

[46] Wang YL, Zhao XH, Du JH, Chen Q. Study on pressure sensitivity properties of cement-based composites with nano-sized carbon black. Bulletn Chin Ceram Soc 2009;28(1):189–93.

[47] Patel JP. Broader use of steel slag aggregates in concrete. Bachelor of science in civil engineering. India: Maharaja Sayajirao University of Baroda; 2006.

[48] Shi CJ, Krivenko PV, Roy DM. Alkali-activated cements and concretes. Taylor and Francis; 2006.

[49] http://www.phxslag.com/phxmgt_region_ne.html.

[50] http://www.advancedmaterialscouncil.org/materials/material_review.php?t=category&id=2&material_id=408.

[51] http://www.chem.ox.ac.uk/icl/heyes/structure_of_solids/lecture1/lec1.html.

[52] http://www.graphite.co.jp/image/sem/rp-l.jpg.

[53] Liu XM, Wu SP. Research on the conductive asphalt concrete's piezoresistivity effect and its mechanism. Constr Build Mater 2009;23(8):2752–6.

[54] Zhang J, Niu HT, Zhou JG, Wang XG, Lin T. Synergistic effects of PEK-C/VGCNF composite nanofibres on a trifunctional epoxy resin. Compos Sci Technol 2011;71:1060–7.

[55] http://www.sigmaaldrich.com/china-mainland/zh/materials-science/nanomaterials/carbon-nanofibers.html.

[56] http://www.chem.wisc.edu.

[57] Gao D, Sturm M, Mo YL. Electrical resistance of carbon-nanofiber concrete. Smart Mater Struct 2009;18:095039.

[58] Han BG, Zhang K, Yu X, Kwon E, Ou JP. Fabrication of piezoresistive CNT/CNF cementitious composites with superplasticizer as dispersant. J Mater Civ Eng 2012;24(6):658–65.

[59] Zheng Z, Feldman D. Synthetic fiber-reinforced Concrete. Prog Polym Sci 1995;20:185–210.

[60] Baeza FJ, Galao O, Zornoza E, Garcés P. Effect of aspect ratio on strain sensing capacity of carbon fiber reinforced cement composites. Mater Des 2013;51:1085–94.

[61] Garas VY, Vipulanandan C. Destructive and non-destructive evaluation of carbon fiber reinforced cement mortar (CFRC). CIGMAT-2004 Conference and Exhibition; 2004. p. 1–3.

[62] Zhou ZJ, Xiao ZG, Pan W, Xie ZP, Luo XX, Jin L. Carbon-coated-nylon-fiber-reinforced cement composites as an intrinsically smart concrete for damage assessment during dynamic loading. J Material Sci Technol 2003;19(6):583–6.

[63] Li GY, Wang PM, Zhao XH. Mechanical behavior and microstructure of cement composites incorporating surface-treated multi-walled carbon nanotubes. Carbon 2005;43:1239–45.

[64] Lourie O, Cox DE, Wagner HD. Buckling and collapse of embedded carbon nanotubes. Phys Rev Lett 1998;81:1638.

[65] Thess A, Lee R, Nikolaev P, Dai HJ, Petit P, Robert J, et al. Crystalline ropes of metallic carbon nanotubes. Science 1996;273:483–7.

[66] Sanchez F, Ince C. Microstructure and macroscopic properties of hybrid carbon nanofiber/silica fume cement composites. Compos Sci Technol 2009;69:1310–8.

[67] Wang C, Li KZ, Li HJ, Xu GZ. CVI treatment of short carbon fibers and their dispersion in CFRC. Acta Mater Compos Sin 2007;24(1):135–40.

[68] Fu XL, Chung DDL. Effect of curing age on the self-monitoring behavior of carbon fiber reinforced mortar. Cem Concr Res 1997;27(9):1313–8.

[69] Banfill PFG, Starrs G, Derruau G, McCarter WJ, Chrisp TM. Rheology of low carbon fiber content reinforced cement mortar. Cem Concr Compos 2006;28:773–80.

[70] Han BG, Guan XC, Ou JP. Experimental research of electrical conductivity and pressure-sensitivity of carbon fiber reinforced cement. Material Sci Technol 2006;14(1):1–4.

Chapter 3

Processing
of Self-Sensing
Concrete

Chapter Outline

3.1 Introduction and Synopsis

Self-sensing concrete is a composite containing a variety of compositions, including matrix composition materials with unique chemical binding capability, functional fillers with complex morphologies and high surface energy, and

45

Self-Sensing Concrete in Smart Structures. http://dx.doi.org/10.1016/B978-0-12-800517-0.00003-4

dispersion materials with specific surface physicochemical properties. This endows the self-sensing concrete mixture with complex thermodynamic and dynamic characteristics. To obtain self-sensing concrete with stable and reproducible properties, an effective processing technology needs to be adopted for incorporating each component into the composites. Generally, processing for self-sensing cement concrete includes three steps: mixing/dispersing, molding, and curing, as shown in Figure 3.1 [1,2].

Processing of self-sensing asphalt concrete is different from that of self-sensing cement concrete. Generally, processing for self-sensing asphalt concrete is as follows: (1) the asphalt is heated until it flows fully; (2) the functional fillers are mixed with heated asphalt binder; (3) other components are blended with the mixtures; and (4) the final mixture is compacted for molding [3,4]. Processing for self-sensing polymer concrete is basically the same as that for self-sensing cement concrete: (1) the functional fillers and aggregate are added slowly into polymer and mixed long enough to obtain a uniform mixture; and (2) the mixture is molded and cured [5]. Because previous research mainly focused on self-sensing cement concrete, this chapter will introduce its processing and the effect of processing on the structures and properties of the composites.

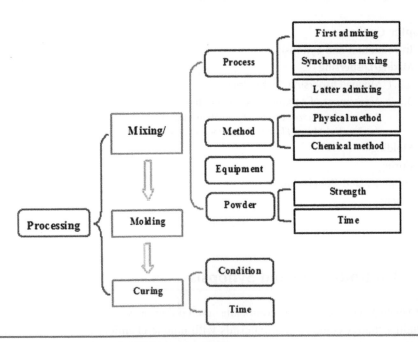

Figure 3.1 Processing of self-sensing concrete.

3.2 Mixing/Dispersing

3.2.1 MIXING/DISPERSING TECHNOLOGY

For self-sensing concrete, mixing equals dispersing. It includes the dispersion of functional fillers in concrete matrix and the dispersion of aggregates in cement paste. Mixing/dispersing is the most important step in composite processing, which greatly affects the uniformity and sensing behavior of self-sensing concrete. Ideally, functional fillers should be uniformly dispersed in the matrix to form an effective electrical network (as shown in Figure 3.2) [6]. The process, method, equipment, and energy adopted in mixing/dispersing are four important aspects influencing the effectiveness of dispersion. According to the order in which functional fillers are added, mixing/dispersing processes in preparing self-sensing concrete can be classified into three groups, as shown in Figure 3.1 and Figure 3.3. They include the first admixing method, the synchronous admixing method, and the latter admixing method. Suitable mixing/dispersing processes for different functional fillers are summarized in Table 3.1 [6–19].

These mixing/dispersing processes should be used jointly for hybrid types of functional fillers. For example, Lin et al. first adopted the synchronous admixing method to mix/disperse carbon black and then adopted the latter admixing method to mix/disperse polyvinyl alcohol fiber when they prepared self-sensing concrete with hybrid carbon black and polyvinyl alcohol fiber [19]. Ou and Han mixed/dispersed carbon fiber by using the first admixing method and then mixed/dispersed carbon black using the synchronous admixing method when fabricating th self-sensing concrete with hybrid carbon fiber and carbon black [20]. Fan first mixed/dispersed graphite powder using the synchronous admixing method and then mixed/dispersed

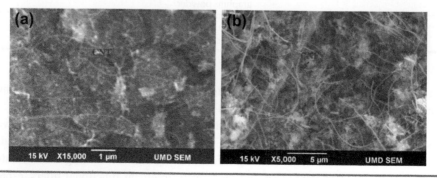

Figure 3.2 Scanning electron microscopy photos for self-sensing concrete [6], (a) with carbon nanotube (CNT), (b) with carbon nanofiber (CNF).

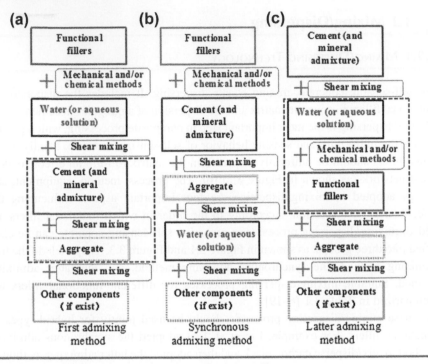

Figure 3.3 Three mixing/dispersing processes for self-sensing concrete.

TABLE 3.1 Suitable Mixing/Dispersing Processes for Different Functional Fillers

Mixing/Dispersing Technology	Suitable Functional Fillers
First admixing method	Carbon fiber, carbon nanotube, carbon nanofiber, carbon black, nano TiO_2, nano Fe_2O_3
Synchronous admixing method	Carbon fiber, steel fiber, steel slag, magnetic fly ash, graphite powder
Latter admixing method	Carbon black, steel fiber, nickel powder, graphite powder, polyvinyl alcohol fiber

carbon fiber by using the latter admixing method when fabricating self-sensing concrete with hybrid graphite powder and carbon fiber [21].

Existing mixing/dispersing methods for functional fillers can be divided into two basic categories: physical and chemical, as shown in Figure 3.1. Physical methods (i.e., mechanical methods) such as high shear mixing and ball milling can separate functional fillers from each other, but they easily fragment functional fillers, especially fibrous fillers. This will decrease the aspect ratios of functional fillers. Also, mechanical mixing methods are difficult for uniformly dispersing nano-scale functional fillers. Ultrasonication is another commonly used physical method to disperse functional fillers into a mixed-water aqueous solution. It can avoid damage to fillers and is especially suitable for nano-scale fillers. Chemical methods (i.e., surface modification methods) are designed to alter the surface structures of functional fillers, either noncovalent surface modification by surfactants or covalent surface modification. They are commonly used to improve the wettability of functional filler surfaces, thus enhancing the solubility and dispersibility of functional fillers [22]. The covalent surface modification method uses the surface functionalization of functional fillers to improve their chemical compatibility with the target media, i.e., to improve their wettability and reduce their tendency to agglomerate. However, aggressive chemical functionalization, such as the use of strong acids at high temperature, might introduce structural defects and result in inferior properties for the functional fillers (e.g., decrease in mechanical property, electrical property, aspect ratio) [22]. The noncovalent surface modification method is particularly attractive because it causes almost no impairment to pristine functional fillers, has low energy consumption, and has controllability [23]. For some functional fillers (e.g., carbon fiber, carbon nanotube and carbon nanofiber), it is difficult for the use of a physical or chemical method alone to separate them [24], so different physical and/or chemical methods are often used jointly to disperse these functional fillers.

Conventional concrete shear mixers such as forced action mixers and rotating concrete mixers (as shown in Figure 3.4 [25]) are the most widely used

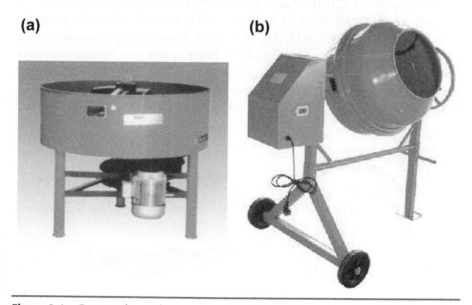

(a)   **(b)**

Figure 3.4 Commonly used concrete mixers [25], (a) forced action mixer (b) rotating concrete mixer.

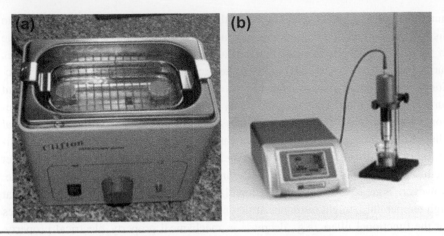

Figure 3.5   (a) Bath-type and (b) probe-type ultrasonicators [26].

mixing/dispersing equipment for self-sensing concrete. Ultrasonicators (including bath and probe ultrasonicators, as shown in Figure 3.5 [26]) and ball mills (as shown in Figure 3.6 [26]) [27] are also often used as mixing/dispersing equipments in some research work, but they need to be used in combination with conventional concrete shear mixers. For example, ultrasonicators often assist in the dispersion of functional fillers in water or an aqueous solution with surfactant in the first admixing method (as shown in Figure 3.7 [2]) [6,23,28]. The ball mill is often used to improve the dispersion of functional fillers in a cement and/or mineral admixture using the synchronous admixing method [26,27]. In addition, proper mixing/dispersing energy is required to effectively disperse functional fillers in concrete. Mixing/dispersing energy is determined by the mixing/dispersing intensity and time [15,28,29]. Several researchers have observed that mixing/dispersing intensity and time will affect the uniformity of the self-sensing concrete composite to a high degree [2,29]. Two

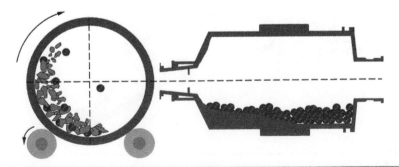

Figure 3.6   Schematics of ball milling technique [26].

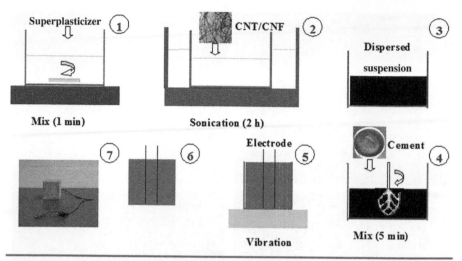

Figure 3.7 Fabrication process of self-sensing concrete with carbon nanotube (CNT)/nanofiber (CNF)[6].

conflicting effects should be balanced during the mixing/dispersing of self-sensing concrete. To ensure that functional fillers are dispersed uniformly throughout the concrete matrix, self-sensing concrete should be stirred sufficiently, which means that high mixing/dispersing energy (e.g., severe and long-time stirring) will produce a positive effect on the improvement of uniformity of the composites. On the other hand, however, high mixing/dispersing energy will generate greater shearing forces and lead to heavy breakage of some functional fillers.

3.2.2 ASSESSMENT OF MIXING/DISPERSING EFFECTIVENESS

Two types of assessment methods are often used to evaluate the mixing/dispersing effectiveness of functional fillers in concrete: the microstructure observation method and the macro electrical resistance measurement method.

3.2.2.1 The Microstructure Observation Method

Observing the microstructure of self-sensing concrete is a commonly used method to evaluate the mixing/dispersing uniformity of composites (as shown in Figure 3.8). Observation equipment often used includes a scanning electron microscope (SEM) and optical microscope. Generally, the microstructure observation method can give only a qualitative description of mixing/dispersing uniformity [2,29–32]. It is also difficult to access the overall dispersion of function fillers with microstructure observation because it samples the composite in only a limited area.

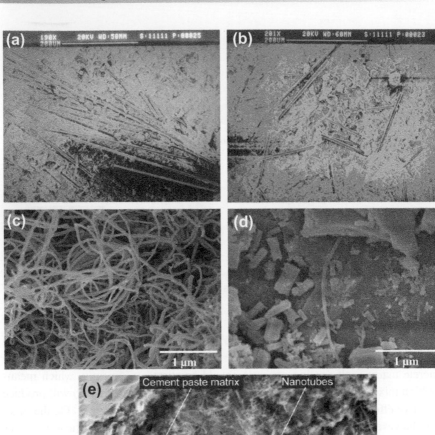

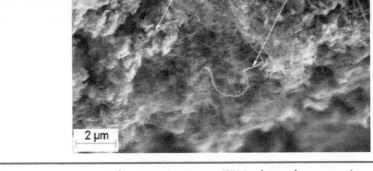

Figure 3.8 Scanning electron microscopy (SEM) photos for comparison of functional filler dispersion, (a) poor dispersion of carbon fiber [2], (b) good dispersion of carbon fiber [2], (c) poor dispersion of carbon nanotube [30] (d) good dispersion of carbon nanotube [30], (e) SEM images of concrete with carbon nanotube [31].

3.2.2.2 The Macro Electrical Resistance Measurement Method

Functional fillers form a three-dimensional conductive network in the concrete matrix. As a result, the electrical resistance of the electrically conductive concrete depends on the conductive network inside the composite. The dispersion of functional fillers in a concrete matrix will directly affect the conductive network inside the composite, and thus the electrical resistance of the composite and its discreteness. Uniformly dispersed functional fillers more easily form a conductive network, presenting better electrical conductivity, smaller electrical resistance discreteness, and a more reproducible sensing property. Therefore, the electrical resistance, electrical resistance discreteness, and sensing property reproducibility of composites can be taken as quantitative indicators to evaluate the mixing/dispersing uniformity of composites (as shown in Figure 3.9 and 3.10) [2,32–34].

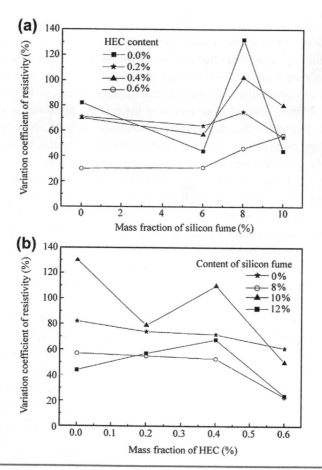

Figure 3.9 Variation coefficient of electrical resistivity of carbon fiber concrete with different contents of hydroxyethyl cellulose (HEC) and silicon fume [33].

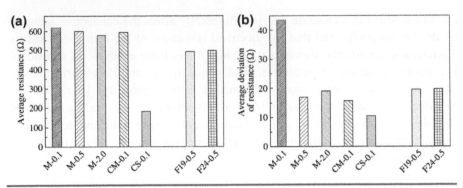

Figure 3.10 Resistance for seven types of carbon nanotube/nanofiber concrete: (a) average resistance; (b) average deviation of resistance [6].

Beside the two methods mentioned previously, other methods have been used to evaluate the mixing/dispersing effectiveness of specific functional fillers in concrete. For example, Yang used the fresh mixture method to access the mixing/dispersing effectiveness of carbon fiber in concrete matrix. The process of the fresh mixture method is as follows: first get some pieces of the mixture of about the same weight from different areas of fresh concrete with carbon fiber, wash the fibers clean, dry and weigh them, and then calculate the content of carbon fiber. Finally, use the calculated standard deviation of the weight of carbon fiber to determine the mixing/dispersing effectiveness of carbon fibers in the concrete matrix [32]. Many researchers have observed the dispersion of functional fillers including carbon fiber, carbon nanotube, and carbon nanofiber in aqueous solution to evaluate the effectiveness of mixing/dispersing (as shown in Figure 3.11 [6] and Figure 3.12 [31]). The dispersion of these functional fillers in aqueous solution is analyzed through transmission electron microscope observation, laser particle size analyzer measurement, SEM observation, atomic force microscope observation, dynamic light scattering measurement, and so forth. [23,24,30,35].

The mixing/dispersing effectiveness of functional fillers in concrete not only depends on mixing/dispersing technology, but is also closely related to the materials, such as the filler type, filler concentration, filler morphology, surfactant type, surfactant concentration, and water-to-cement ratio. Commonly, a low filler concentration, large filler size and low aspect ratio, hydrophilic filler surface, high surfactant concentration, and high water-to-cement ratio are beneficial for enhancing the mixing/dispersing effectiveness of functional fillers in concrete [2,23,36–39]. For example, Saafi et al. studied the effect of carbon nanotube concentration on their dispersion in concrete. The experimental results showed that the carbon nanotubes were uniformly distributed within the geopolymer cement concrete matrix at 0.1 and 0.5 wt.% of carbon nanotube concentration (as shown in

Figure 3.11   Observation of carbon nanotube suspension dispersed with superplasticizer [6].

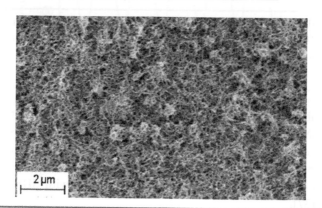

Figure 3.12   Scanning electron microscopy image of water solution of carbon nanotube [31].

Figure 3.13(a) and (b)) and were poorly distributed and severely agglomerated within the matrix at 1 wt.% of carbon nanotube concentration (as shown in Figure 3.13(c)) [36].

Han et al. used sodium dodecyl sulfate (SDS) and sodium dodecyl benzene sulfonate (NaDDBS) as surfactants to improve the dispersion of multi-walled

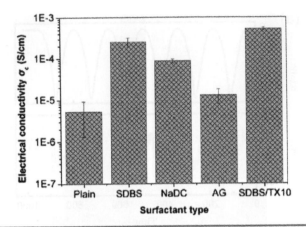

Figure 3.15   Electrical conductivity (mean value and deviation) of plain cement paste and carbon nanotube cement pastes using SDBS, NaDC, AG, and SDBS and Triton X-100 as surfactants, respectively [38].

prepared using casting and vibration technology presented the largest fractional change amplitude of conductance, whereas that prepared by pressing technology had the smallest fractional change amplitude of conductance (as shown in Figure 3.16). This is because extrusion and the pressing technologies could improve pore structure distribution insides composites (as shown in Figure 3.17 and Figure 3.18). The decrease in porosity results in the improvement of the elastic modulus of the

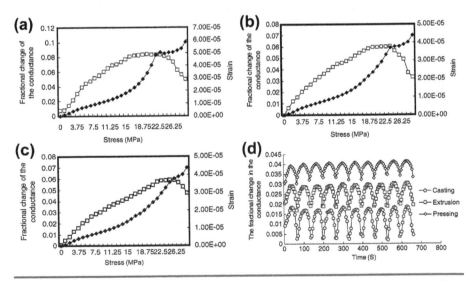

Figure 3.16   Fractional changes of conductance of self-sensing composite prepared using different fabrication methods [41], (a) casting, (b) extrusion, (c) pressing, (d) under repeated compressive loading.

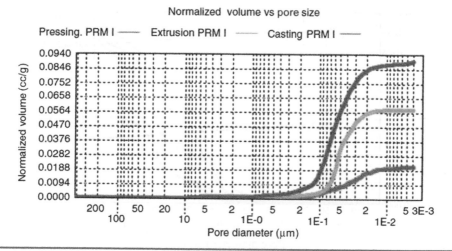

Figure 3.17 Pore distribution using different pore diameters (PRM) of different self-sensing concretes [41].

composites. Under the same stress, the larger the modulus of the composites, the smaller the strain is of the composites. As a result, composites prepared by casting and vibration technology have the largest maximum change in electrical resistivity and the highest sensing sensitivity. In addition, the sensing properties of the composites prepared by extrusion and pressing technologies have better repeatability and stability compared with that of composites prepared using casting and vibration technology. This is attributed to improvement in the interface structure between carbon fiber and concrete matrix during the extrusion and pressing processes [41].

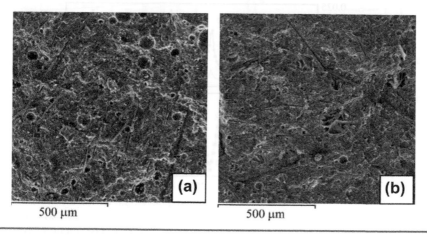

Figure 3.18 Scanning electron microscopy photos of carbon fiber concrete with different molding technologies [42], (a) casting, (b) extrusion.

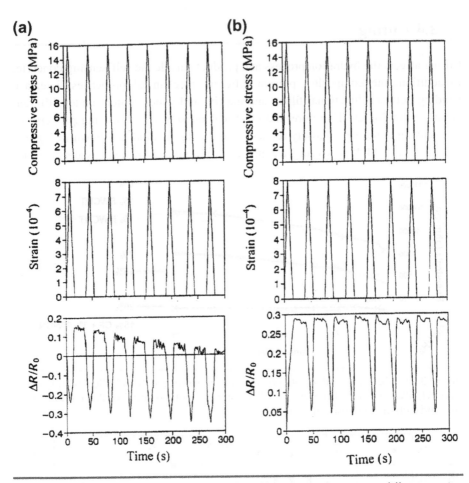

Figure 3.23 Self-sensing behavior of carbon fiber cement mortar at different curing ages [45], (a) at 28 days of curing, (b) at 7 days of curing.

condition and curing age. In previous research, curing conditions such as standard curing (i.e., at $20 \pm 3\,°C$ and above 90% relative humidity or in water), hot water curing, and steam curing have been used to prepare self-sensing concrete [7,16,40]. Jia investigated the effect of three curing methods (i.e., 28 days of water curing, 7 days of water curing and 21 days of air curing, and 1 day of $90\,°C$ hot water curing and 27 days of water curing) on the sensing behavior of concrete with steel slag and/or magnetic fly ash. He observed that the sensitivies of the three types of composites decreased in order. The increase in curing age from 1 to 28 days led to a decrease in sensitivity of the composites, whereas the maximum change in electrical resistivity increased because the ultimate strength of the composites increased with curing age [16]. Fu and Chung studied the sensing behavior of concrete with carbon fiber at curing age from 7 to 28 days (as shown in Figure 3.23). The electrical resistance increased monotonically with compressive strain during the first loading at 7 days, but at 14 and 28 days it decreased monotonically. This was due to weakening of the fiber–cement interface as curing progressed. Electrical resistance decreased slightly and irreversibly at the end of each cycle as cycling progressed at 14 and 28 days, but not at 7 days. This resulted from the increasing rigidity of the cement matrix as curing progressed [45].

3.5 Summary and Conclusions

Self-sensing concrete is a multicomponent composite system with complex thermodynamic and dynamic characteristics. An effective processing technology is needed to incorporate each component material into the composites to form self-sensing concrete with desirable structure and properties. Commonly, the processing technology of self-sensing concrete includes three steps: mixing/dispersing, molding, and curing. Mixing/dispersing is the most important step in composite processing, which would greatly affect the distribution of functional fillers in a concrete matrix. The mixing/dispersing effectiveness of functional fillers in concrete can be evaluated using the microstructure observation method, macro electrical resistance measurement method, fresh mixture method, and observation method of aqueous solution of functional fillers. Molding of self-sensing concrete determines its compaction. Four molding technologies, including casting and vibration, extrusion, pressing, and hydrothermal hot-pressing technologies, have been used to fabricate self-sensing concrete. Curing of self-sensing concrete includes two aspects, condition and age, which dominate pore characteristics, hydration products, and water content inside the composites. Variety in self-sensing concrete structures induced by processing technology has a strong effect on its mechanical and electrical performances. Processing technology thus is a key issue to be considered for fabricating self-sensing concrete, such as the components of the composites.

References

[1] Mehta PK, Monteiro PJM. Concrete: microstructure, properties and materials. New York: McGraw-Hill; 2006.

[2] Han BG. Properties, sensors and structures of pressure-sensitive carbon fiber cement paste [dissertation for the Doctor Degree in Engineering]. China: Harbin Institute of Technology; 2005.

[3] Wu SP, Mo LT, Shui ZH, Chen Z. Investigation of the conductivity of asphalt concrete containing conductive fillers. Carbon 2005;43:1358–63.

[4] Liu XM, Wu SP, Ye QS, Qiu J, Li B. Properties evaluation of asphalt-based composites with graphite and mine powders. Constr Build Mater 2008;22:121–6.

[5] Sett K. Characterization and modeling of structural and self-monitoring behavior of fiber reinforced polymer concrete [dissertation for the Master of Science in Civil Engineering]. USA: University of Houston; 2003.

[6] Han BG, Zhang K, Yu X, Kwon E, Ou JP. Fabrication of piezoresistive CNT/CNF cementitious composites with superplasticizer as dispersant. J Mater Civ Eng 2012;24(6):658–65.

[7] Chung DDL. Self-monitoring structural materials. Mater Sci Eng R Rep 1998;22(2):57–78.

[8] Hou TC, Lynch JP. Conductivity-based strain monitoring and damage characterization of fiber reinforced cementitious structural components. Proc SPIE 2005;5765:419–29.

[9] Mao QZ, Zhao BY, Sheng DR, Li ZQ. Resistance changement of compression sensible cement speciment under different stresses. J Wuhan Univ Technol 1996;11:41–5.

[10] Li H, Xiao HG, Ou JP. A study on mechanical and pressure-sensitive properties of cement mortar with nanophase materials. Cem Concr Res 2004;34(3):435–8.

[11] Li H, Xiao HG, Ou JP. Effect of compressive strain on resistance of carbon black filled cement-based composites. Cem Concr Compos 2006;28:824–8.

[12] Li CT, Qian JS, Tang ZQ. Study on properties of smart concrete with steel slag. China Concr Cem Prod 2005;2:5–8.

[13] Han BG, Yu X, Kwon E. A self-sensing carbon nanotube/cement composite for traffic monitoring. Nanotechnology 2009;20:445501 (5 pp).

[14] Gao D, Sturm M, Mo YL. Electrical resistance of carbon-nanofiber concrete. Smart Mater Struct 2009;18:095039.

[15] Azhari F. Cement-based sensors for structural health monitoring [dissertation for the Master Degree of Applied Science]. Canada: University of British Columbia; 2008.

[16] Jia XW. Electrical conductivity and smart properties of fe1-σO waste mortar [dissertation for the Doctor Degree in Engineering]. China: Chongqing University; 2009.

[17] Hong L. Study on the smart properties of the graphite slurry infiltrated steel fiber concrete [dissertation for the Doctor Degree in Engineering]. China: Dalian University of Technology; 2006.

[18] Han BG, Han BZ, Ou JP. Experimental study on use of nickel powder-filled cement-based composite for fabrication of piezoresistive sensors with high sensitivity. Sens Actuators A Phys 2009;149(1):51–5.

[19] Lin VWJ, Mo L, Lynch JP, Li VC. Mechanical and electrical characterization of self-sensing carbon black ECC. Proc SPIE Int Soc Opt Eng 2011;7983:798316 (12 pp.).

[20] Ou JP, Han BG. Piezoresistive cement-based strain sensors and self-sensing concrete components. J Intell Mater Syst Struct 2009;20(3):329–36.

[21] Fan XM. Study of piezoresistivity of smart concrete and mechanism on reinforcement corrosion and protection [dissertation for the Doctor Degree in Engineering]. China: Wuhan University of Technology; 2009.

[22] Vaisman L, Wagner HD, Marom G. The role of surfactants in dispersion of carbon nanotubes. Adv Colloid Interface Sci 2006;128–130:37–46.

[23] Han BG, Yu X, Ou JP. Chapter 1: multifunctional and smart carbon nanotube reinforced cement-based materials. In: Gopalakrishnan K, Birgisson B, Taylor P, Attoh-Okine NO, editors. Book: nanotechnology in civil infrastructure: a paradigm shift, vol. 1–47. Publisher: Springer; 2011. p. 276.

[24] De Ibarra YS, Gaitero JJ, Campillo I. Atomic force microscopy and anoindentation of cement pastes with nanotube dispersions. Phys Status Solidi A 2006;203:1076–81.

[25] http://dict.youdao.com/wiki/%E6%90%85%E6%8B%8C%E6%9C%BA/#.

[26] Hwang SH, Park YB, Yoon KH, Bang DS. Chapter 18 smart materials and structures based on carbon nanotube composites. In: Yellampalli S, editor. Book: carbon nanotubes-synthesis, characterization, applications. Publisher: InTech; 2011. pp. 371–96. 514 p.

[27] Li GY. A new method to modify the dispersion and properties of carbon nanotube/cement matrix. www.papter.edu.cn:1–6.

[28] Han BG, Guan XC, Ou JP. Application of ultrasound for preparation of carbon fiber cement-based composites. Mater Sci Technol 2009;17(3):368–72.

[29] Shah SP, Konsta-Gdoutos MS, Metexa ZS. Highly-Dispersed Carbon Nanotube-Reinforced Cement-Based Materials. Patent US 2009, 0229494A1.

[30] Konsta-Gdoutos MS, Metaxa ZS, Shah PS. Highly dispersed carbon nanotube reinforced cement based materials. Cem Concr Res 2010;40(7):1052–9.

[31] Materazzi AL, Ubertini F, D'Alessandro A. Carbon nanotube cement-based transducers for dynamic sensing of strain. Cem Concr Compos 2013;37:2–11.

[32] Yang YX. Methods study on dispersion of fibers in CFRC. Cem Concr Res 2002;32:747–50.

[33] Wang C, Li KZ, Li HJ, Xu GZ. CVI treatment of short carbon fibers and their dispersion in CFRC. Acta Mater Compos Sin 2007;24(1):135–40.

[34] Yang YX, Mao QZ, Shen DR, Li ZQ. Study on the dispersion of fiber in carbon fiber cement composites. J Build Mater 2001;4(1):84–8.

[35] Luo JL. Fabrication and functional properties of multi-walled carbon nanotube/cement composites [dissertation for the Doctoral Degree in Engineering]. Harbin, China: Harbin Institute of Technology; 2009.

[36] Saafi M, Andrew K, Tang PL, McGhon D, Taylor S, Rahman M, et al. Multifunctional properties of carbon nanotube/fly ash geopolymeric nanocomposites. Constr Build Mater 2013;49:46–55.

[37] Han BG, Yu X, Ou JP. Dispersion of carbon nanotubes in cement-based composites and its influence on the piezoresistivities of composites. ASME 2009 Conference on Smart Materials, Adaptive Structures and Intelligent Systems (SAMASIS09), Oxnard; 2009.

[38] Luo JL, Duan ZD, Li H. The influence of surfactants on the processing of multi-walled carbon nanotubes in reinforced cement matrix composites. Phys Status Solidi 2009;206(12):2783–90.

[39] Han BG, Yu X, Kwon E, Ou JP. Effects of CNT concentration level and water/cement ratio on the piezoresistivity of CNT/cement composites. J Compos Mater 2012;46(1):19–25.

[40] Wang XF, Wang YL, Jin ZH. Electrical conductivity characterization and variation of carbon fiber reinforced cement composite. J Mater Sci 2002;37:223–7.

[41] Cheng X, Wang SD, Lu LC, Huang SF. Influence of preparation process on piezo-conductance effect of carbon fiber sulfoaluminate cement composite. J Compos Mater 2011;45(20):2033–7.

[42] Cheng X, Wang SD, Huang SF, Chen W. Influence of forming technology on piezoresistivity of CFSC. J Build Mater 2008;11(1):84–8.

[43] Huang SF, Xu DY, Xu RH, Chang J, Lu LC, Cheng X. Microcosmic and smart properties of carbon fiber cement-based composites. Acta Mater Compos Sin 2006;23(4):95–9.

[44] Chen B, Wu KR, Yao W. Conductivity of carbon fiber reinforced cement-based composites. Cem Concr Compos 2004;26:291–7.

[45] Fu XL, Chung DDL. Effect of curing age on the self-monitoring behavior of carbon fiber reinforced mortar. Cem Concr Res 1997;27(9):1313–8.

Chapter 4

Measurement of Sensing Signal of Self-Sensing Concrete

Chapter Outline

4.1 Introduction and Synopsis

The measurement of the sensing property of self-sensing concrete is as crucial as its processing. The sensing property of self-sensing concrete is the coupled relation between electrical and mechanical properties. The mechanical properties of **67**

Self-Sensing Concrete in Smart Structures. http://dx.doi.org/10.1016/B978-0-12-800517-0.00004-6

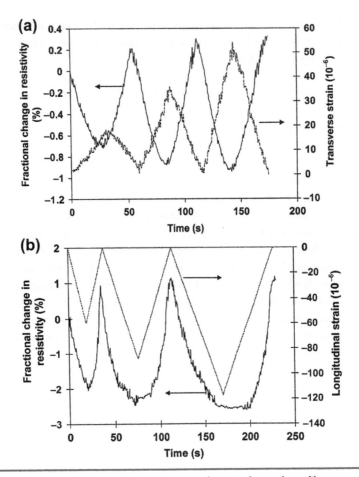

Figure 4.2 Fractional change in resistivity and strain for carbon fiber cement paste. (a) Transverse direction. (b) Longitudinal direction [10].

have can be used as electrical signals to describe the sensing behavior of self-sensing concrete (as shown in Figures 4.5 and 4.6) [12].

4.2.2 IMPEDANCE OR ELECTRICAL REACTANCE

Under AC rather than DC conditions, the impedance Z consists of the electrical resistance R (real part of Z) and the electrical reactance X (imaginary part of Z), i.e.:

$$Z = R + iX \tag{4.4}$$

Zheng et al. and Fu et al. found that the impedance and the electrical reactance are more sensitive indicators for strain than the electrical resistance (as shown in Figures 4.7 and 4.8) [13].

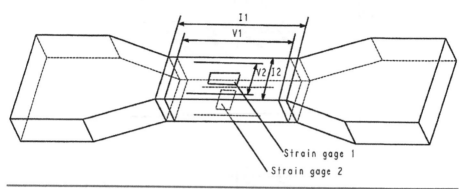

Figure 4.3 Sample configuration for measuring the transverse electrical resistivity during uniaxial tension. (I 1 and V 1 indicate current and voltage contacts, respectively, for longitudinal resistance measurement; I 2 and V 2 indicate current and voltage contacts, respectively, for resistance measurement.) [11]

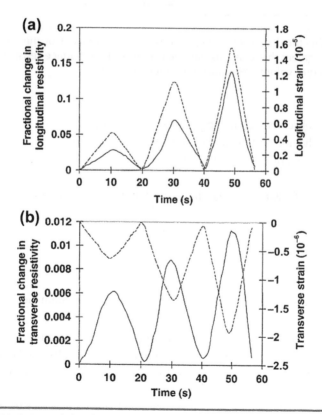

Figure 4.4 Variation of the fractional change in transverse electrical resistivity with time (solid curve) and of the strain with time (dashed curve) during dynamic uniaxial tensile loading at increasing stress amplitudes within the elastic regime for carbon fiber cement paste. (a) Longitudinal direction. (b) Transverse direction [11].

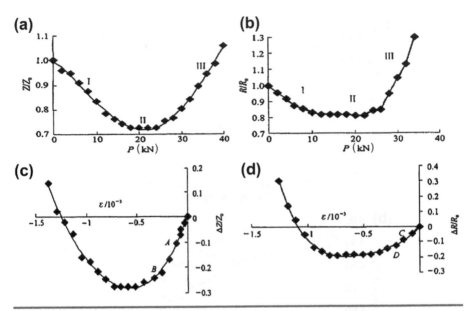

Figure 4.7   Sensing property of cement paste with carbon fiber during static compressive testing to failure. (a) Relationship between impedance (Z) and compressive force (P). (b) Relationship between change in electrical resistance (R/R_0) and compressive force (P). (c) Relationship between impedance (Z) and compressive strain (ε). (d) Relationship between change in electrical resistance (R/R_0) and compressive strain (ε) [13].

4.3.1 ELECTRODE MATERIALS

Electrode materials should have two basic features in nature: low electrical resistance and stable electrically conductive property. By now, the materials used as electrodes of self-sensing concrete mainly include metal flake (e.g., copper, stainless steel , and lead flake) with and without a hole (as shown in Figure 4.12) [20,21]; metal foil (e.g., copper, stainless steel, and aluminum foil); metal mesh (e.g., copper and stainless steel mesh); copper loop; metal bar (e.g., copper and stainless steel bar); carbon rod; copper tape; copper wire; steel wool; graphite cloth; and conductive paint (e.g., silver, copper, graphite, and carbon black) [15–24]. Han et al. [25] and Azhari [26] observed that electrode materials have a strong effect on measured electrical resistance (as shown in Figure 4.13) [26]. Figures 4.14 and 4.15 show the effect of electrode materials on the current density distribution [27].

4.3.2 ELECTRODE FIXING AND LAYOUT

The electrode materials mentioned above separately or jointly serve as electrodes of self-sensing concrete in the attachment (as shown in Figures 4.12(b) and 4.16),

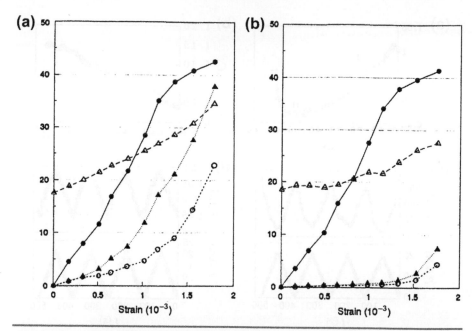

Figure 4.8 Sensing property of cement mortar with carbon fiber during static compressive testing to failure (the compressive stress in MPa (solid circles), fractional increase in electrical resistance (open circles), fractional increase in electrical reactance (solid triangles)). (a) Cement mortar with methylcellulose, silica fume, and fibers at 7 days. (b) Cement mortar with latex and fibers at 7 days [14].

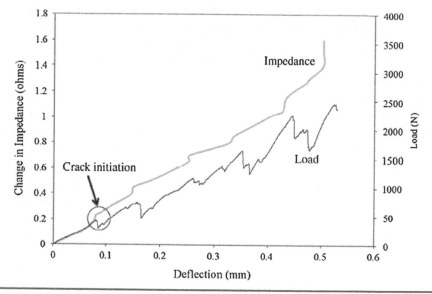

Figure 4.9 Change in impedance vs. deflection response of geopolymer cement concrete with 0.5 wt.% of carbon nanotubes under bending [15].

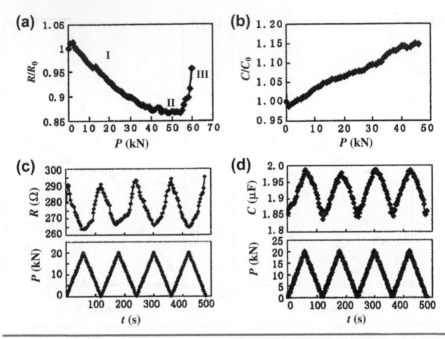

Figure 4.10 Sensing property of concrete with carbon fiber. (a) Relationship between electrical resistance and compressive force under uniaxial loading. (b) Relationship between capacitance and compressive force under uniaxial loading. (c) Relationship between electrical resistance and compressive force under repeated loading. (d) Relationship between capacitance and compressive force under repeated loading [16].

embedment (as shown in Figure 4.12(d) and (e)), and plating or clipping (as shown in Figure 4.17) fixing styles [26–29]. Among these fixing styles, attachment and embedment are most widely used [20–29]. In addition, the electrodes of self-sensing concrete are usually set in the two-probe layout (as shown in Figures 4.12(c) and 4.14) or the four-probe layout (as shown in Figures 4.12(e) and 4.13) [12,19,20,29–42].

The previous experimental and simulation results indicate that the fixing style and layout of electrodes have a direct bearing on the electrical field distribution in the self-sensing concrete, thereby affecting measurement results of electrical resistivity [19,27]. Currently, there are six easily accessible schemes of electrode fixing and layout, as shown in Figure 4.18. The electrodes are attached on surface of the self-sensing concrete in schemes (a) and (b) [34,42]. The two schemes have been widely used in laboratory measurement of the sensing property of self-sensing concrete because they do not impair the mechanical property of the self-sensing concrete. However, the attached electrodes will easily be debonded from the self-sensing concrete in practical application. To overcome this issue, four schemes (as shown in (c)–(f) in Figure 4.18) are proposed. The embedded mesh, perforated plate or loop electrode can minimize the effect of electrode incorporation, and ensure

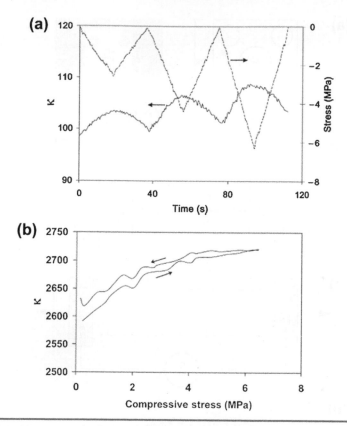

Figure 4.11   Variation of relative dielectric constant (κ) during repeated application of compressive stress. (a) For carbon fiber cement paste. (b) For steel fiber cement paste [18].

the integrity of electrode and the self-sensing concrete. Moreover, the self-sensing concrete can provide protection to the embedded electrodes [19,20,27,32,43].

It should be noted that the shape of self-sensing concrete specimens is set as a cube or cuboid subjected to compression in Figure 4.18, and no tension or bending specimen is considered. Actually, the fixing styles and layouts of electrodes mentioned above are also effective for other shapes of specimens and even for tensile specimens. When the shape or the loading mode of specimens changes, only some small adjustments are needed.

4.4 Measurement Method of Electrical Resistance

Corresponding to the layouts of electrodes, the electrical resistance measurement methods of self-sensing concrete include the two-probe method (as shown in

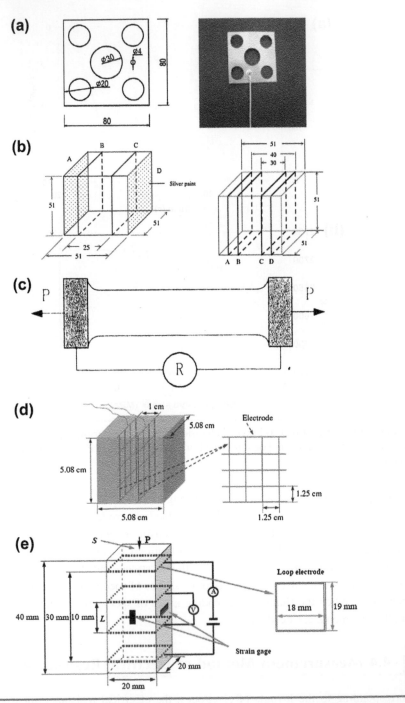

Figure 4.12   Typical electrodes for self-sensing concrete. (a) Metal flake with hole as electrode [20]. (b) Silver paint in conjunction with copper wire as electrode [21]. (c) Silver paint as electrode [22]. (d) Stainless steel mesh as electrode [23]. (e) Copper loop as electrode [24].

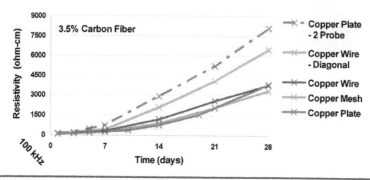

Figure 4.13   Effect of electrode on the measured resistivity of concrete with carbon fiber by using four-probe method [26].

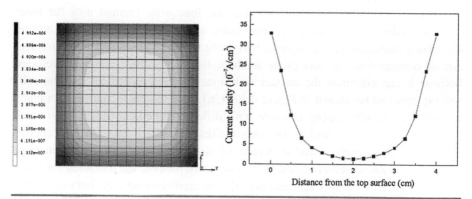

Figure 4.14   Current density distribution of a loop electrode [27].

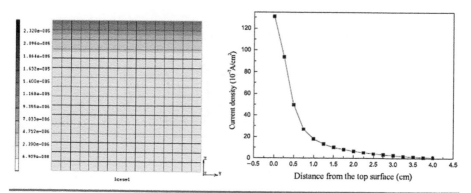

Figure 4.15   Current density distribution of a line electrode [27].

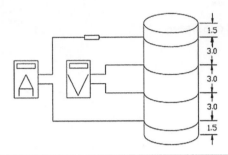

Figure 4.16 Attachment fixing style of electrode [28].

Figures 4.12(c) and 4.17), three-probe method, and four-probe method (as shown in Figures 4.12(d) and 4.16). The three-probe method is rarely used due to complex electrode setting [44]. In the two-probe method, the two electrodes are both current electrode and voltage electrodes, whereas the four-probe method uses the inner two electrodes and the outer two electrodes are voltage electrodes and current electrodes, respectively. Although the two-probe method a has simpler measurement circuit compared to the four-probe method, the four-probe method is preferred because it can eliminate the contact resistance between electrodes and the self-sensing concrete (as shown in Figure 4.19) [26]. This has been confirmed in many experiments on self-sensing concrete with different functional fillers. Han et al. proposed a method to calculate the contact resistance, which found that the contact resistance is much higher than the electrical resistance of self-sensing concrete [19]. Chiarello and Zinno observed that, in the two-probe method, the measured electrical resistance value is strongly influenced by the electrode contact area, but in the four-probe method the electrical resistance is almost independent [39]. Li [32] and Jia [45] found that the initial electrical resistivity and strain sensitivity coefficient of self-sensing concrete measured by the two-probe method varies with the size of

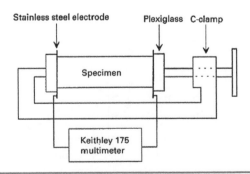

Figure 4.17 Clipping fixing style of electrode [29].

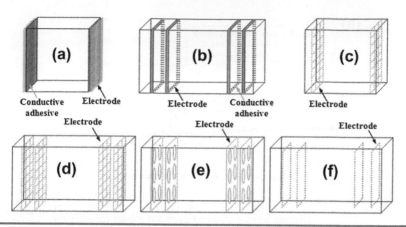

Figure 4.18 The commonly used fixing style and layout of electrodes in self-sensing concrete.

measured specimen and the space between electrodes. This is because the contact resistance occupies different percentages in the initial electrical resistances of specimens with different sizes and different electrode spaces. Chung suggested that the two-probe method gives less effective and less reliable sensing than the four-probe method, due to the inclusion of contact resistance [46].

Based on the previous research results, the four-probe method is recommended for electrical resistance measurement of self-sensing concrete. However, some researchers pointed out that although the measured resistance value from the two-probe method may be higher than the true resistance of the composites due to the influence of the contact resistance, it does not affect the capability of the two-probe method to detect changes in resistance of the composites under loading [47]. In addition, the two-probe method is simpler and more convenient to use compared to the four-probe method. Therefore, it is still widely used in research and application of self-sensing concrete.

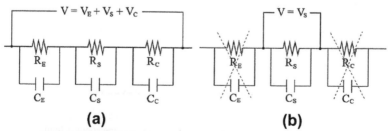

V_E, R_E and C_E: voltage, resistance and capacitance from the electrodes, respectively;
V_S, R_S and C_S: voltage, resistance and capacitance from the sensor, respectively;
V_C, R_C and C_C: voltage, resistance and capacitance from the contact, respectively.

Figure 4.19 Equivalent circuits for (a) two-probe method and (b) four-probe method [26].

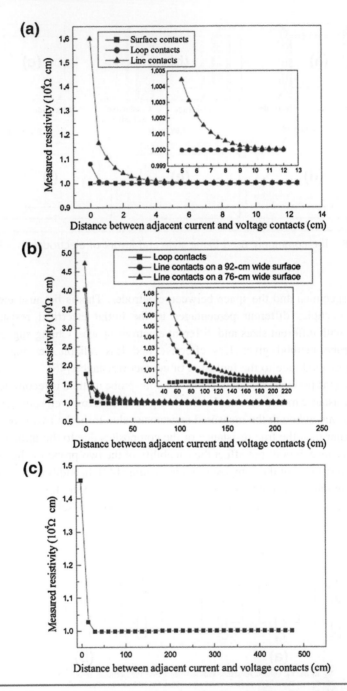

Figure 4.20 Dependence of the measured resistivity of different concrete specimens obtained by simulation on the distance between the adjacent voltage and current contacts. (a) Beam-shaped specimen. (b) Column-shaped specimen. (c) Slab-shaped specimen [27].

It should be noted that the measurement results of electrical resistance may be relative to some electrode configuration parameters in the four-probe method. Han et al. observed that the area of current electrodes has an effect on the measured electrical resistivity, but the measured electrical resistivity is not affected by the area of voltage electrodes and the mesh size of electrodes. The spacing between the current and voltage electrodes also influences the measured electrical resistivity before exceeding a critical value [19]. Zhu and Chung verified that the measured electrical resistivity in the four-probe method decreases with increase of the spacing between the adjacent voltage and current electrodes through a finite element simulation (as shown in Figure 4.20) [27].

In measuring the electrical resistance of self-sensing concrete by using either the four-probe method or the two-probe method, the direct current (DC) test method is the simplest way, with a fixed voltage applied to the tested specimen. However, as a constant electrical field is applied during the electrical resistance measurement, the movement and aggregation of the ions in the cement concrete matrix will lead to an electrical polarization in the self-sensing cement concrete, the electrical conduction of which is dominated by the ionic conduction mechanism (as shown in Figure 4.21) (the detailed conduction mechanism will be discussed in Chapter 6). Due to an exponential rise presenting in the measured resistance caused by the polarization effect (as shown in Figure 4.22(a)) [33], it is difficult to accurately measure the changes in electrical resistance of the self-sensing concrete caused by external loading with the DC measurement. One way to nullify the polarization effect in the DC electrical resistance measurement is to record the change in resistance caused by polarization

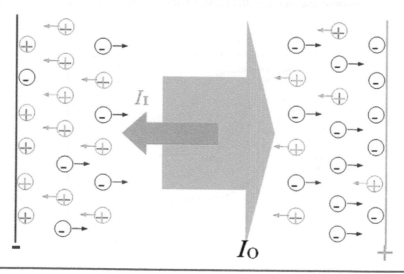

Figure 4.21   Sketch of polarization in concrete.

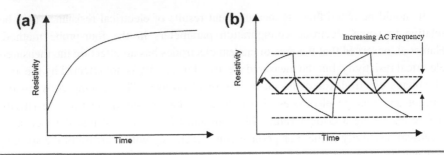

Figure 4.22   Polarization effect when applied with (a) DC and (b) AC [33].

for an unloaded specimen. Another approach is to apply a DC voltage potential well ahead of loading the composite so as to allow the resistance to plateau off after a complete polarization. However, both methods are not practical for use in the field. Furthermore, polarization is dependent upon the specimen geometry, with large specimens taking longer time to be fully polarized. An alternative method is to employ alternating current (AC) signals with equal magnitudes of positive and negative peaks to the self-sensing cement concrete. Although polarization can still be observed in the AC signals, its effect is lessened to an acceptable range by increasing frequency of the applied AC voltage (as shown in Figures 4.22(b) and 4.23) [33,48–50]. Fu et al. [14] and Zheng et al. [13] compared the sensing behavior of the carbon fiber cement mortar and cement paste measured with the DC and the AC methods, respectively. They observed that the AC impedance measurement may also reflect the whole deformation process of the composites when they are compressed. The sensing property measured with the AC method is more stable, and the sensitivity

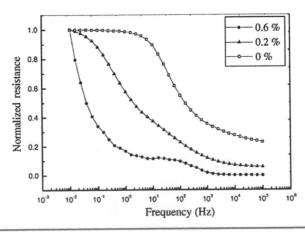

Figure 4.23   Change in electrical with increasing frequency of AC test signal [50].

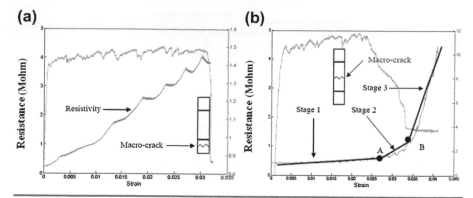

Figure 4.24   Stress–stain and resistivity–strain relationship of a carbon-ECC during AC monotonic load testing. (a) Specimen in which major cracking occurs beyond the probes. (b) Specimen in which cracking occurs within the measuring region [33].

is higher. Hou and Lynch tested the sensing property of the engineered cementitious composites with hybrid polyvinyl alcohol fiber and carbon fiber respectively with the DC and the AC methods. They observed that unlike the DC method, the AC resistance test has little or no polarization influencing the conductivity measurement, thus resulting in strain measurements with greater accuracy (as shown in Figure 4.24) [33]. Moreover, Han et al. explored the effect of AC voltage on the sensing behavior of the carbon nanotube/cement composites. They observed that a low-amplitude AC voltage is helpful for improving the pressure-sensitive sensitivity of the composites [49].

4.5 Acquisition and Processing of Sensing Signal

In the two-probe method, the electrical resistance signal can be directly collected using multimeters, or indirectly collected using a Wheatstone bridge method (as shown in Figure 4.25) [51]. In the DC four-probe method, which needs

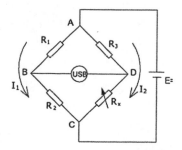

Figure 4.25   Schematic diagram of the electrical resistance measurement using a Wheatstone bridge [51].

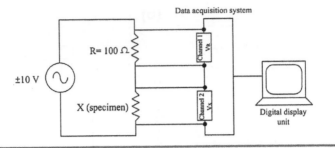

Figure 4.26 Simplified circuit diagram for electrical resistivity measurement [52].

synchronously collecting current and voltage signals, the resistance is calculated from voltage divided by current. The voltage signal can be directly collected by using an A/D data acquisition card or multimeter, but the current signal is difficult to collect synchronously. To avoid this issue, a simple method is to connect a constant reference resistance in series with the self-sensing concrete. In this way, to measure the voltage signal at both ends of the reference resistance is equivalent to measuring the current signal, because the current signal can be obtained by dividing the reference resistance into the collected voltage signal at both ends. Then the electrical resistance of the self-sensing concrete can be calculated from dividing the voltage signals at both ends of the self-sensing concrete by the current. This indicates that the electrical resistance of self-sensing concrete can be obtained through synchronously collecting voltage signals at both ends and at both ends of the reference resistance [19]. This method, which transforms measurement of an electrical resistance signal into measurement of an electrical voltage signal, can also be used in the two-probe method (as shown in Figures 4.26 and 4.27) [26,52]. In addition, the electrical resistance signal can also be

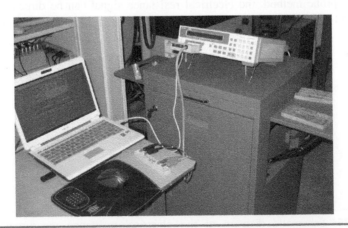

Figure 4.27 Example of an electrical resistivity test setup [26].

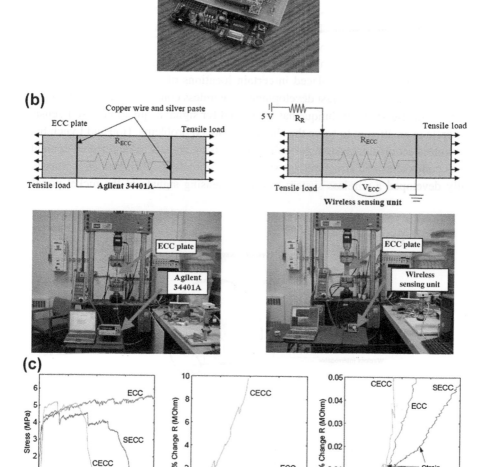

Figure 4.28   Wireless acquisition system of sensing signal of the engineered cementitious composites. (a) Wireless active sensing unit prototype. (b) Experimental setup. (c) Test results of wireless system [33, 56].

and high-pass filtering. In this method, the high-pass filter can eliminate the noise signal caused by the polarization effect (as shown in Figure 4.30) [49].

4.6 Summary and Conclusions

The characterization of self-sensing concrete is a key problem in its research and application. It has a directly relation with the sensing signal measurement. Five types of electrical parameters (including electrical resistance or resistivity, impedance, electrical reactance, capacitance, and relative dielectric constant) have been taken as sensing signals to characterize sensing property of self-sensing concrete. The direct-current volume electrical resistivity or resistance is the most common electrical parameter because it is fully able to describe the sensing behavior of self-sensing concrete.

The electrode fabrication and the electrical resistance measurement methods determine the accuracy of sensing signal of self-sensing concrete. Fabrication of the electrodes for self-sensing concrete involves three aspects: choice of electrode materials, fixing of electrodes, and layout of electrodes. Corresponding to the layouts of electrodes, the electrical resistance measurement methods of self-sensing concrete mainly include the two-probe method and the four-probe method. The four-probe method is recommended for electrical resistance measurement of self-sensing concrete. However, the two-probe method is still widely used because it is simpler and more convenient to use compared to the four-probe method.

The electrical signals of self-sensing concrete are conventionally acquired using a wired acquisition method. The wireless measurement method has been used for signal acquisition of self-sensing concrete to address some problems of the wired acquisition method. In general, the acquired sensing signals will inevitably be polluted by measurement noise. The subsequent signal processing technology is beneficial for removing the measurement noise and extracting true sensing signal information.

References

[1] Shi ML. AC impedance spectroscopy principles and applications. Beijing: National Defense Industry Press; 2001.
[2] Wansom S, Kidner NJ, Woo LY, Mason TO. AC-impedance response of multi-walled carbon nanotube/cement composites. Cem Concr Compos 2006;26:509–19.
[3] Zhang X, Ding XZ, Ong CK, Tan BTG, Yang J. Dielectric and electrical properties of ordinary Portland cement and slag cement in the early hydration period. J Mater Sci 1996;31:1345–52.
[4] Tuan CY. Electrical resistance heating of conductive concrete containing steel fibers and shavings. ACI Mater J 2004;101(1):65–71.

[5] Farrar JJ. Electrically conductive concrete. GEC J Sci Technology 1978;45(1):45–8.

[6] Whittington H, McCarter W, Forde MC. Conduction of electricity through concrete. Mag Concr Res 1981;33(114):48–60.

[7] Chen PW, Chung DDL. Carbon fiber reinforced concrete as a smart material capable of non-destructive flaw detection. Smart Mater Struct 1993;2:22–30.

[8] Mao Q, Sun M, Chen P. Study on volume resistance and surface resistance of CFRC. J Wuhan Univ Technol 1997;19(2):65–7.

[9] Kovacs G. Micromachined transducers sourcebook. Boston: McGraw-Hill; 1998.

[10] Wen SH, Chung DDL. Uniaxial compression in carbon fiber-reinforced cement, sensed by electrical resistivity measurement in longitudinal and transverse directions. Cem Concr Res 2001;31:297–301.

[11] Wen SH, Chung DDL. Uniaxial tension in carbon fiber reinforced cement, sensed by electrical resistivity measurement in longitudinal and transverse directions. Cem Concr Res 2000;30:1289–94.

[12] Meehan DG, Wang SK, Chung DDL. Electrical-resistance-based sensing of impact damage in carbon fiber reinforced cement-based materials. J Intell Mater Syst Struct 2010;21(1):83–105.

[13] Zheng LX, Song XH, Li ZQ. Investigation on the method of AC measurement of compression sensivility of carbon fiber cement. J Huazhong Univ Sci Technol (Urban Sci Ed) 2005;22(2):27–9.

[14] Saafi M, Andrew K, Tang PL, McGhon D, Taylor S, Rahman M, et al. Multifunctional properties of carbon nanotube/fly ash geopolymeric nanocomposites. Constr Build Mater 2013;49:46–55.

[15] Fu XL, Ma E, Chung DDL, Anderson WA. Self-monitoring in carbon fiber reinforced mortar by reactance measurement. Cem Concr Res 1997;27(6):845–52.

[16] Zheng LX, Song XH, Li ZQ. Study on the compression sensibility of CFRC under quasi-triaxial compression. Bull Chin Ceram Soc 2004;4:40–3.

[17] Wang SD, Huang SF, Chen W, Cheng X. Smart properties of carbon fiber reinforced sulphoaluminate cement. Acta Mater Compos Sin 2005;22(6):114–9.

[18] Wen SH, Chung DDL. Cement-based materials for stress sensing by dielectric measurement. Cem Concr Res 2002;32(9):1429–33.

[19] Han BG, Guan XC, Ou JP. Electrode design, measuring method and data acquisition system of carbon fiber cement paste piezoresistive sensors. Sensor Actuator Phys 2007;135(2):360–9.

[20] Hong L. Study on the smart properties of the graphite slurry infiltrated steel fiber concrete [dissertation for the doctor degree in engineering]. China: Dalian University of Technology; 2006.

[21] Wen SH, Chung DDL. The role of electronic and ionic conduction in the electrical conductivity of carbon fiber reinforced cement. Carbon 2006;44:2130–8.

[22] Zhou ZJ, Xiao ZG, Pan W, Xie ZP, Luo XX, Jin L. Carbon-coated-nylon-fiber-reinforced cement composites as an intrinsically smart concrete for damage assessment during dynamic loading. J Mater Sci Technol 2003;19(6):583–6.

[23] Han BG, Yu X, Kwon E, Ou JP. Piezoresistive MWNTs filled cement-based composites. Sens Lett 2010;8(2):344–8.

[24] Han BG, Han BZ, Ou JP. Novel piezoresistive composite with high sensitivity to stress/strain. Mater Sci Technol 2010;26(7):865–70.

[25] Han BG, Han BZ, Ou JP. Experimental study on use of nickel powder-filled cement-based composite for fabrication of piezoresistive sensors with high sensitivity. Sensor Actuator Phys 2009;149(1):51–5.

[26] Azhari F. Cement-based sensors for structural health monitoring [dissertation for the master degree of applied science]. Canada: University of British Columbia; 2008.

[27] Zhu SR, Chung DDL. Numerical assessment of the methods of measurement of the electrical resistance in carbon fiber reinforced cement. Smart Mater Struct 2007;16:1164–70.

[51] Wang W, Wu SP, Dai HZ. Fatigue behavior and life prediction of carbon fiber reinforced concrete under cyclic flexural loading. Mater Sci Eng a 2006;431:242–51.

[52] Chacko RM. Carbon fiber reinforced cement-based sensors [dissertation for the master degree of applied science]. Canada: University of British Columbia;2005.

[53] Xiang LX. Study on the compression sensibility of carbon concrete and its structure [dissertation for the master degree in science]. Wuhan University of Technology; 2004.

[54] Han BG, Yu Y, Han BZ, Ou JP. Development of a wireless stress/strain measurement system integrated with pressure-sensitive nickel powder-filled cement-based sensors. Sens Actuators A 2008;147:536–43.

[55] Shen SB, Kidd S, Feng Z, et al. A sensor-based road traffic condition sensing[M] using the surface infrastructure. Sensor Tr. America J. 2005;1 (no)1–no(no).

[56] Han TG, Lynch JP. Monitoring strain in engineered cementitious composites using wireless sensors. In: Proceedings of the international conference on fracture (ICF-XII), Ottawa, Canada, March 20–23, 2009, pp.1–6.

[57] Han BG, Zhang K, Yu X, Kwon E, Ou JP. Nickel phosphate-based self-sensing pavement for vehicle detection. Measurement 2011;44(9):1645–50.

Sensing Properties of Self-Sensing Concrete

Chapter Outline

Self-Sensing Concrete in Smart Structures. http://dx.doi.org/10.1016/B978-0-12-800517-0.00005-8

▌5.1 Introduction and Synopsis

The sensing behavior of self-sensing concrete can be described by the relationship between fractional change in electrical resistance and external force (or stress and strain) as follows:

$$\frac{\Delta\rho}{\rho_0} = f(X) \tag{5.1}$$

where X may be external force F, stress σ, or strain ε.

Sensitivity, which is an important factor in evaluating the sensing property of self-sensing concrete, can be characterized by such parameters as the maximum amplitude of fractional change in electrical resistance ($\max(|\Delta\rho/\rho_0|)$), force sensitivity coefficient ($(\Delta\rho/\rho_0)/F$), stress sensitivity coefficient ($(\Delta\rho/\rho_0)/\sigma$), or strain sensitivity coefficient ($(\Delta\rho/\rho_0)/\varepsilon$, also called gage factor). For example, $\max(|\Delta\rho/\rho_0|)$ of self-sensing concrete with carbon fiber, steel slag, carbon black, and nickel powder can reach 45%, 35%, 30%, and 80%, respectively, under compression. The $(\Delta\rho/\rho_0)/\sigma$ of self-sensing concrete with carbon fiber and nickel powder can reach 0.028/MPa and 0.05/MPa, respectively, under compression. The gage factor of self-sensing concrete with carbon fiber, steel fiber, and nickel powder can reach 560, 720, and 895, respectively, under compression, while the gage factor of self-sensing concrete with carbon fiber, steel fiber, and nickel powder can reach 90, 4560, and 8584, respectively, under tension. In addition, some other parameters such as input/output range, linearity, repeatability, hysteresis, signal-to-noise ratio, and zero shifts are also used to assist in characterizing the sensing property of self-sensing concrete [1–3]. This chapter will introduce the sensing characteristics of self-sensing concrete under different loading conditions, and the effect of some factors on the sensing property of self-sensing concrete.

▌5.2 Sensing Characteristics under Different Loading Conditions

Self-sensing concrete features different sensing behaviors when it is subjected to different loads such as compression, tension, and flexure, which will be introduced in the following sections.

5.2.1 UNDER COMPRESSION

5.2.1.1 Monotonic Compression

As shown in the lower part of Figure 5.1, when under compression, the $\Delta\rho/\rho_0$ of the self-sensing concrete usually goes through the states of decrease, balance, and

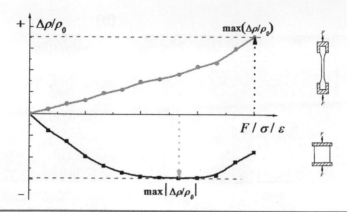

Figure 5.1 Typical sensing behavior of self-sensing concrete under compression and tension.

abrupt increase, which correspond respectively to the pressure compaction, the germination of fresh cracks, and the extension of crack under monotonic uniaxial compression. The pressure compaction causes functional fillers to approach each other, thus improving the conductive network inside the composite. The germination of fresh cracks leads to destruction and reconstruction of the conductive network inside the composite. The extension of cracks results in breakdown of the conductive network [4].

This phenomenon has been confirmed in many subsequent experimental studies on self-sensing concrete with different fillers under monotonic uniaxial compression (as shown in Figures 5.2–5.8), although the patterns of relationship curve between electrical resistivity and compressive loading (or stress and strain) have some difference (as shown in Figures 5.3–5.8) [4–12]. In addition, $\Delta\rho/\rho_0$ was found to be monotonously increased upon compression up to fracture in some research work. This phenomenon is also considered as a sensing property [13–15].

Besides uniaxial compression, some researchers investigated the sensing property of self-sensing concrete under monotonic multiaxial compression. For example, Wu et al. compared the stress-sensing behaviors of carbon fiber concrete under monotonic uniaxial compression and monotonic biaxial compression. They observed that the sensing behavior curve under biaxial compression is similar to that under uniaxial compression, but the stress level at which the fractional change in resistance starts to increase is different. The sensing property of the composite under biaxial compression is more sensitive than that under uniaxial compression (as shown in Figure 5.9) [16].

Zheng et al. tested the sensing property of the carbon fiber cement paste under monotonic quasi-triaxial compression (i.e., composites are loaded at one direction and restrained at the other two directions) and compared it with the result tested

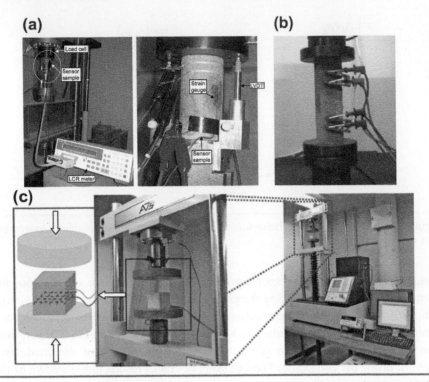

Figure 5.2 Compressive loading setup. (a) Cylindrical specimen [5]. (b) Rectangular specimen [6]. (c) Cubic specimen [7].

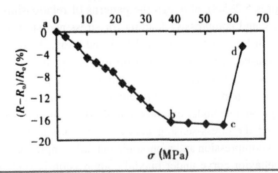

Figure 5.3 Sensing property of concrete with steel slag under monotonic compression [8].

under uniaxial compression. The comparative result shows that the change trends of electrical resistivity with pressure under the two load conditions are almost the same, but the sensitivity reduces and the linearity becomes better under the quasi-triaxial compression [17]. Jia et al. compared the sensing property of the steel slag

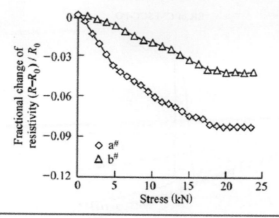

Figure 5.4 Sensing property of concrete with carbon fiber under monotonic compression [9].

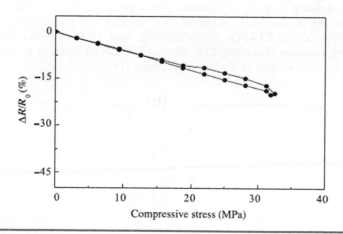

Figure 5.5 Sensing property of concrete with nano Fe_2O_3 under monotonic compression [10].

concrete under monotonic uniaxial compression and monotonic quasi-triaxial compression. They found that the $\max|\Delta\rho/\rho_0|$ is the same in both cases, but the sensing property under monotonic quasi-triaxial compression has a smaller stress sensitivity coefficient and better linearity [18].

5.2.1.2 Repeated Compression

Under repeated compression, the $\Delta\rho/\rho_0$ of self-sensing concrete decreases upon loading and increases upon unloading in each cycle. However, for different

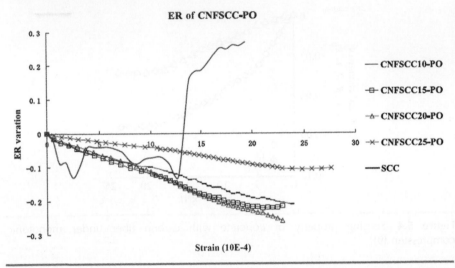

Figure 5.6 Sensing property of concrete with carbon nanofiber under monotonic compression (ER is electrical resistance; SCC denotes self-consolidating concrete; CNFSCC10-PO, CNFSCC15-PO, CNFSCC20-PO, and CNFSCC25-PO denote self-consolidating concrete containing CNF PR-19-XT-PS-OX in amounts of 1.0%, 1.5%, 2.0%, and 2.5% by volume of binder, respectively) [11].

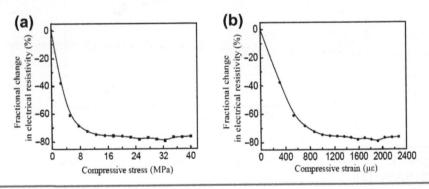

Figure 5.7 Sensing property of concrete with nickel powder under monotonic compression. (a) Relationship between stress and electrical resistivity. (b) Relationship between strain and electrical resistivity. [12].

amplitudes of compressive stresses, the variation of baseline electrical resistivity and the change in electrical resistivity are different. Here the baseline electrical resistivity refers to the electrical resistivity as compressive stress returns to zero under cyclic loading. It is different from the initial electrical resistivity of the specimen never subjected to external force. This is because some internal defects (i.e., native cracks, holes, etc.) in the composites are gradually reduced in cycles of compressive loading. As a result, the electrical resistivity of the composites at the

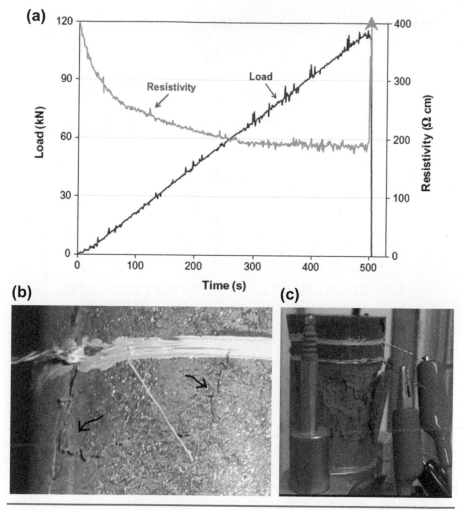

Figure 5.8 Concrete with carbon fiber and carbon nanotube under monotonic compression. (a) Sensing property. (b) Appearance of cracks. (c) Failure [5].

beginning of each cycle is not equal to the initial electrical resistivity, and it is defined as the baseline electrical resistivity. The electrical polarization under DC current will also lead to the change of electrical resistivity with respect to time.

Variation rules of the electrical resistivity of the self-sensing concrete under repeated compression have been obtained from many experiments. They can be summarized as follows. When the compressive stress amplitude is below about 30% of specimen ultimate strength, both the baseline electrical resistivity and the change in electrical resistivity are reversible, which results from the reversible elastic deformation of the composites. When the compressive stress amplitude is the range

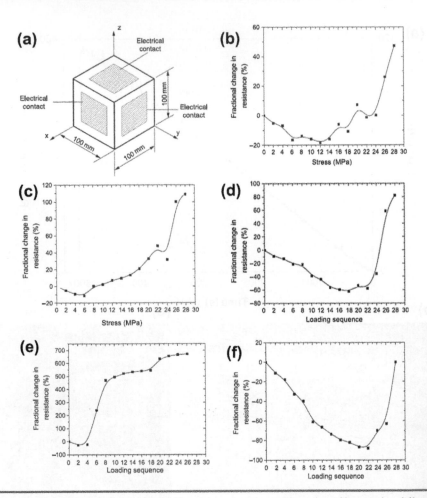

Figure 5.9   Sensing behaviors of self-sensing concrete with carbon fiber under different load modes. (a) Specimen configuration for measuring the electrical resistance in the x, y, and z directions during uniaxial/biaxial compression. (b) Fractional change in resistance in the x direction vs compressive stress in the x direction during uniaxial compression. (c) Fractional change in resistance in the z direction vs compressive stress in the x direction during uniaxial compression. (d) Fractional change in resistance in the x direction vs the loading sequence during biaxial compression. (e) Fractional change in resistance in the y direction vs the loading sequence during biaxial compression. (f) Fractional change in resistance in the z direction vs the loading sequence during biaxial compression [16].

from about 30% to 75% of specimen ultimate strength, the change in electrical resistivity is reversible, while the baseline electrical resistivity is irreversible. This is attributed to minor matrix damage and consequent reconstruction of the conductive network inside the composites. When the compressive stress amplitude is above about 75% of specimen ultimate strength, both the change in electrical resistivity and the baseline electrical resistivity are irreversible. This is because the matrix damage is extensive and the conductive network inside the composites is broken down.

The above results have been confirmed by observing the sensing behaviors of self-sensing concrete with different fillers under uniaxial compression (as shown in Figures 5.10–5.22) [5,19–26]. Moreover, Wang et al. also tested the sensing

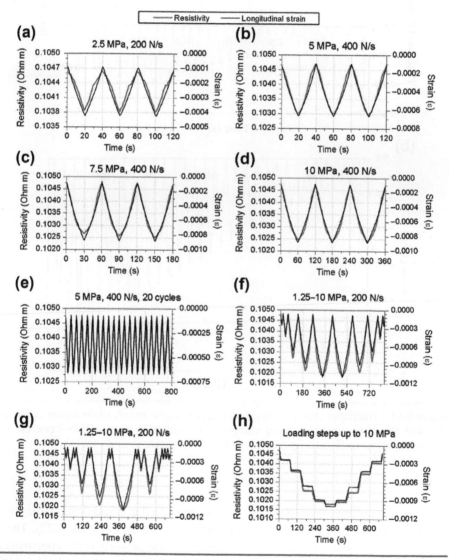

Figure 5.10   Concrete with carbon fiber during repeated compression (the maximum load values and loading rates are indicated in the subtitles of (a) to (h).) [19].

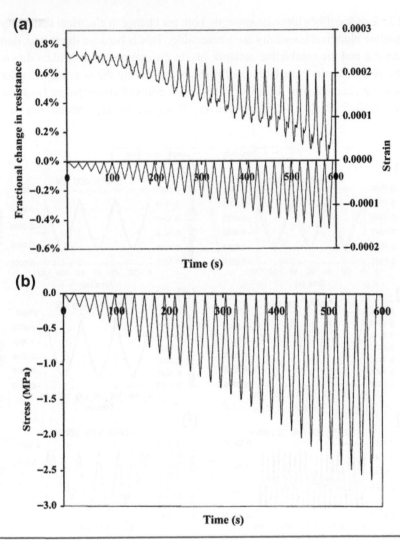

Figure 5.11 Fractional change in resistance (upper curve in (a)), strain (lower curve in (a)), and stress (b) during repeated compressive loading of carbon fiber concrete at increasing stress amplitudes up to 20% of the compressive strength [20].

property of concrete with 0.8% of carbon fiber under a quasi-triaxial cyclic compression with a loading amplitude of 30 kN (as shown in Figure 5.22). They observed that the sensitivity and repeatability of sensing property of the composites in the loading direction under triaxial compression is better than that under uniaxial compression [27].

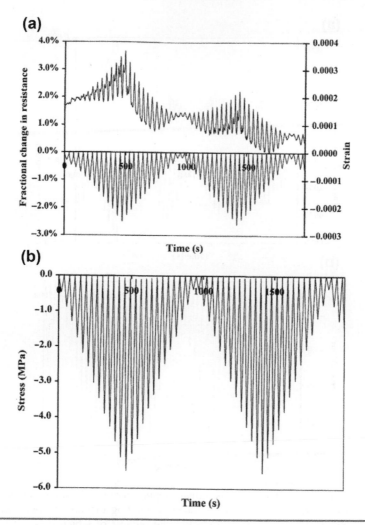

Figure 5.12   Fractional change in resistance (upper curve in (a)), strain (lower curve in (a)), and stress (b) during repeated compressive loading of carbon fiber concrete at increasing and decreasing stress amplitudes, the highest of which was 40% of the compressive strength [20].

5.2.1.3 Impact

The impact load is one of the common loads to concrete structures. However, only a few studies have been conducted on exploring the response of self-sensing concrete to impact load. Jia observed the change in electrical resistance of the $Fe_{1-\sigma}O$ waste mortar under impact loads (as shown in Figure 5.23 [28]). The bigger the amplitude of impact loads, the bigger the change in electrical resistance of the composites.

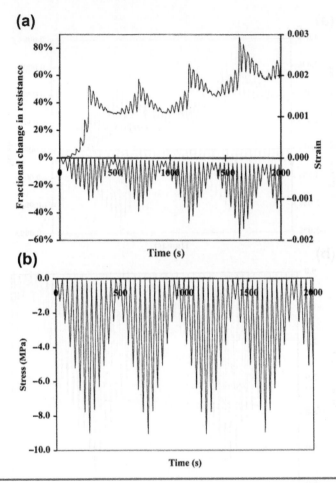

Figure 5.13 Fractional change in resistance (upper curve in (a)), strain (lower curve in (a)), and stress (b) during repeated compressive loading of carbon fiber concrete at increasing and decreasing stress amplitudes, the highest of which was >90% of the compressive strength [20].

Under impact loading of low-amplitude stress, the electrical resistance of the composite after loading is the same as its initial electrical resistance. However, the electrical resistance of the composite after loading is different from its initial electrical resistance under impact loading of high amplitude. This is because damage occurs inside the composite in this case [29].

Meehan et al. tested the damage of self-sensing cement mortar with carbon fiber (as shown in Figure 5.24) when under impact stress. They observed that the surface resistance increases abruptly upon impact due to damage. It decreases abruptly upon impact after a sufficient number of impacts because of the subtle damage of the cement

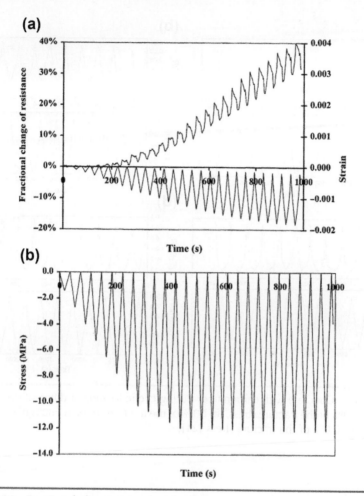

Figure 5.14   Fractional change in resistance (upper curve in (a)), strain (lower curve in (a)), and stress (b) during repeated compressive loading of carbon fiber concrete at increasing stress amplitudes up to >90% of the compressive strength and then with the stress amplitude fixed at the maximum [20].

matrix between adjacent fibers and the consequent increase in the degree of fiber–fiber contact and decrease in the resistance. The surface electrical resistance and impact energy have good corresponding relationship (as shown in Figure 5.25). It is effective for sensing impact damage through the effect of impact on the surface electrical resistance when the region of resistance measurement contains the point of impact [30].

Han et al. found that the impact loads cause the decrease of electrical resistance of self-sensing carbon nanotube cement composite, and the decrease in amplitude of electrical resistance depends on the amplitude of impact loads (as shown in Figure 5.26).

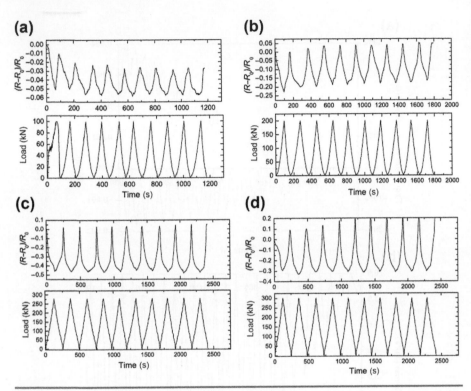

Figure 5.15   Concrete with carbon fiber during cyclic loading at different stress amplitudes: (a) 30% of the fracture stress, (b) 60% of the fracture stress, (c) 80% of the fracture stress, and (d) 0% of the fracture stress [21].

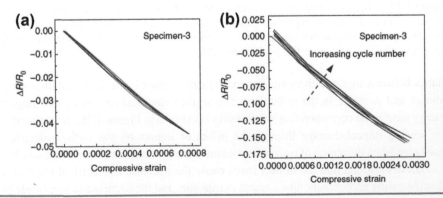

Figure 5.16   Sensing property of self-sensing concrete with carbon black under cyclic loading at different stress amplitudes: (a) 10 kN stress, (b) 35 kN stress [22].

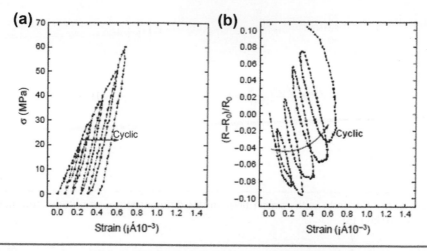

Figure 5.17 Stress–strain–resistance relationship for concrete with carbon fiber subsequent to uniaxial compression cycles with maximum pressures increased step by step [23].

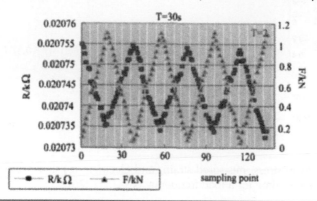

Figure 5.18 Concrete with carbon fiber and graphite during repeated compression [24].

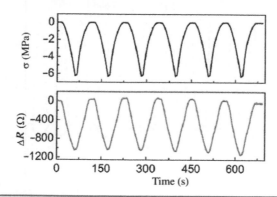

Figure 5.19 Sensing property of self-sensing concrete with carbon nanotube under repeated compressive loading with amplitude of 6 MPa [25].

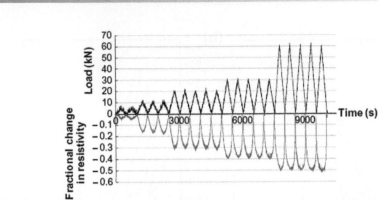

Figure 5.20 Concrete with carbon fiber under different stress amplitudes (ultimate load is 110 kN) [5].

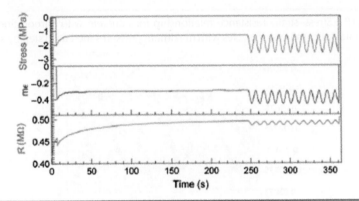

Figure 5.21 Axial stress, measured strain, and electrical resistance of concrete with carbon nanotube vs time for a load frequency of 0.1 Hz [26].

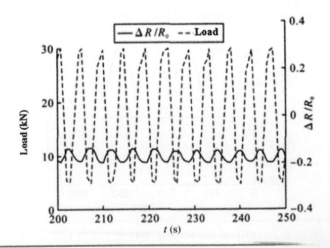

Figure 5.22 Concrete with carbon fiber under triaxial cyclic loading [27].

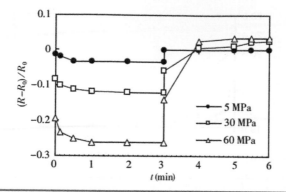

Figure 5.23 Effect of impact on sensing property of concrete with steel slag [28].

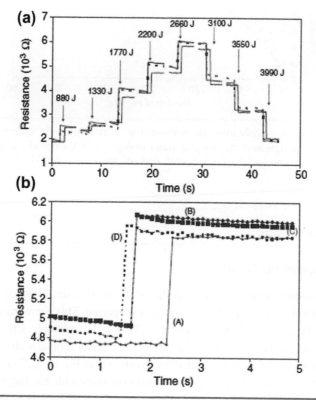

Figure 5.24 (a) Surface resistance upon impact at progressively increasing energy. The impact energy is successively 880, 1330, 1770, 2200, 2660, 3100, 3550, and 3900 J. Two solid curves: AC. Dashed curve: DC. (b) A magnified view of the effect of the 2660 J impact. (A) and (B) are two specimens measured under AC. (C) and (D) are two specimens measured under DC [30].

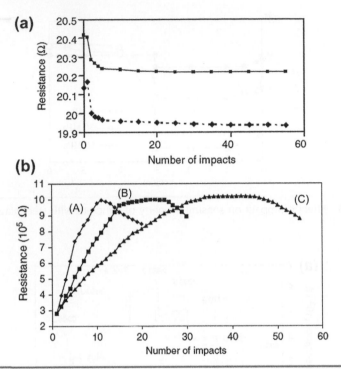

Figure 5.25 Surface resistance of self-sensing concrete with different types of carbon fiber upon repeated impact at a fixed energy. (a) PAN-based carbon fiber and 3990 J of impact energy. (b) Pitch-based carbon fiber: (A) 2200 J, (B) 1770 J, and (C) 1330 J [30].

5.2.2 UNDER TENSION

5.2.2.1 Monotonic Tension

It can be seen from Figure 5.1 that the $\Delta\rho/\rho_0$ of the self-sensing concrete increases in monotonic uniaxial tension as the fillers trend to separate. This phenomenon has been observed in all types of self-sensing concrete [31–40]. For example, Chen and Chung firstly observed that the electrical resistivity of the cement mortar with 0.53 vol.% of carbon fiber increases with tensile loading (as shown in Figure 5.27) [31].

Sett tested the sensing behavior of polymer concretes with 3% and 6% of carbon fiber under tension. The change in electrical resistivity of both polymer concretes increases nonlinearly with tensile stress up to failure [32]. Zhou and Yang investigated the sensing property of the concrete composites with 2.0% of carbon coated nylon fiber under monotonic tension. They observed that the fractional change in electrical resistance of the composites linearly increases with stress and strain. After

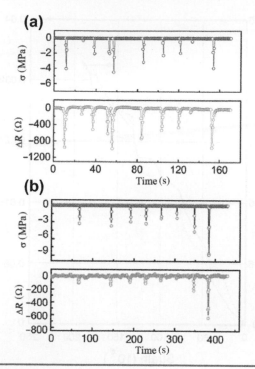

Figure 5.26 Sensing property of self-sensing concrete with carbon nanotube under impulsive loading. (a) Cement paste. (b) Cement mortar [25].

the ultimate tensile strain, the fractional change in electrical resistance changes much faster (as shown in Figure 5.28) [33].

Hou and Lynch investigated the sensing property of engineered cementitious composites with polyvinyl alcohol fiber only, hybrid polyvinyl alcohol fiber and steel fiber, and hybrid polyvinyl alcohol fiber and carbon fiber. They observed that these three types of composites exhibit similar self-sensing properties, i.e., a strong linear relationship between electrical resistivity and mechanical strain. The engineered cementitious composite with 2.0 vol.% of polyvinyl alcohol fiber has a much higher gage factor than those with a hybrid of polyvinyl alcohol fiber and steel fiber (0.1 vol.%) or carbon fiber (0.4 vol.%), although its initial electrical resistivity is higher (as shown in Figure 5.29) [34].

Saafi found that the cement paste with 1.0 vol.% of carbon nanotube exhibits a higher sensitivity to the applied tension stress compared with the cement paste with 0.5 vol.% of carbon nanotube (as shown in Figure 5.30) [35].

Lin et al. observed that an incorporation of carbon black (0.25, 0.5, and 1.0 wt.% by weight ratio to total cementitious content (cement plus fly ash)) is capable of enhancing the electrical resistivity change of the engineered cementitious composite

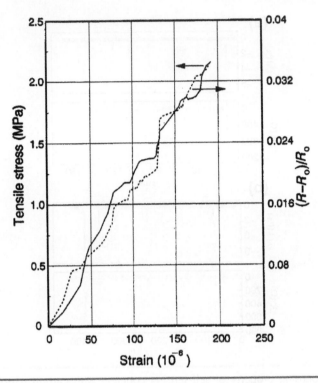

Figure 5.27 Sensing property of concrete with carbon fiber under monotonic tension [31].

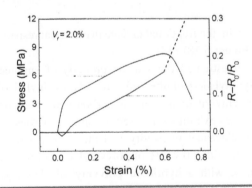

Figure 5.28 Sensing property of concrete with carbon-coated nylon fiber under monotonic tension [33].

with 2.0 vol.% of polyvinyl alcohol fiber. The sensing behavior of the engineered cementitious composite with carbon black not only is a function of strain but also is affected by the cracking behavior of the material when subjected to tension (as shown in Figure 5.31) [36].

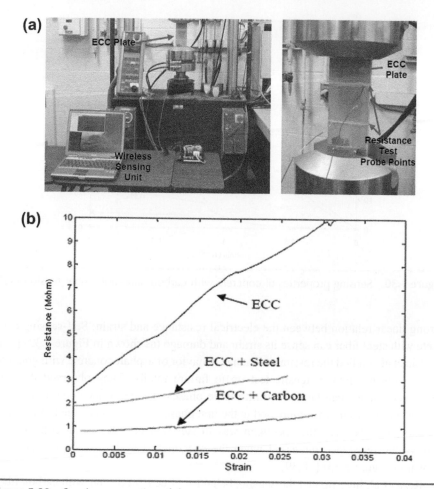

Figure 5.29 Sensing properties of the engineered cementitious composites with polyvinyl alcohol fiber only (ECC), hybrid polyvinyl alcohol fiber and steel fiber (ECC + Steel), and hybrid polyvinyl alcohol fiber and carbon fiber (ECC + Carbon) under tension. (a) Experimental setup. (b) Test results [34].

In addition, Teomete and Kocyigit tested the correlations between the electrical resistance change of self-sensing concrete with 0.2 vol.% of 6 mm steel fiber and tensile strain during a split tensile test. They observed that tensile strain caused fiber elongation and opening of microcracks, which increased electrical resistance. The initiation and propagation of cracks increased the electrical resistance of the self-sensing concrete with steel fiber dramatically. The effect of strain and crack on the electrical resistance change can be decomposed by considering the effect of strain until maximum load. The initiation of a crack starts to decrease the load. The crack has a superior effect on the electrical resistance after the load starts to decrease. There is a

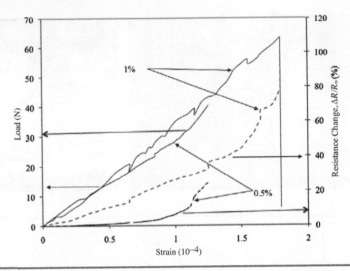

Figure 5.30 Sensing properties of concrete with carbon nanotube under tension [35].

strong linear relation between the electrical resistance and strain. Self-sensing concrete with steel fiber can sense its strain and damage (as shown in Figure 5.32) [37].

Liu et al. studied the resistance change behavior of asphalt concrete with graphite powder during indirect tensile test up to failure totally. They observed that the resistance change can be divided into three phases: The resistance deceased when the specimen was more compressed in the first phase, and then it changed little in the second phase because the specimen was in steady state after being compressed. Finally the resistance increased when the specimen was destroyed gradually (as shown in Figure 5.33) [38,39].

Reza et al. investigated the sensing behavior of cement mortar with carbon fiber in volume percentages from 0% to 0.6% under compact tension. They observed good correlations between the electrical behavior and mechanical behavior of the composite. The electrical resistance can be used to provide some insight on the development and the mechanisms of the fracture process zone, and to provide an estimate of the length of a propagating crack (as shown in Figure 5.34) [40].

5.2.2.2 Repeated Tension

Under repeated tension, the $\Delta\rho/\rho_0$ of self-sensing concrete increases upon loading and decreases upon unloading in each cycle. Like repeated compression, the variation of baseline electrical resistivity and the change in electrical resistivity are different under different amplitudes of tension stress. The baseline electrical resistivity and the change in electrical resistivity are reversible within the elastic region. Under a high amplitude of stress, both change in electrical resistivity and the baseline electrical resistivity are

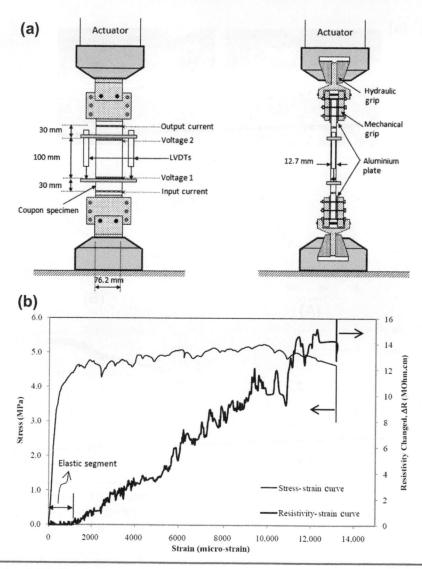

Figure 5.31 Sensing properties of engineered cementitious composites with hybrid polyvinyl alcohol fiber and carbon black under tension. (a) Experimental setup. (b) Test results [36].

irreversible. For example, Chen and Chung [41] and Wen and Chung [42] tested the sensing property of the carbon fiber concrete under repeated tension. They observed that the change in electrical resistance irreversibly increases during the initial portion of the first loading, reversibly increases during the latter portion of the first loading and during any subsequent loading, and reversibly decreases during unloading in any cycle.

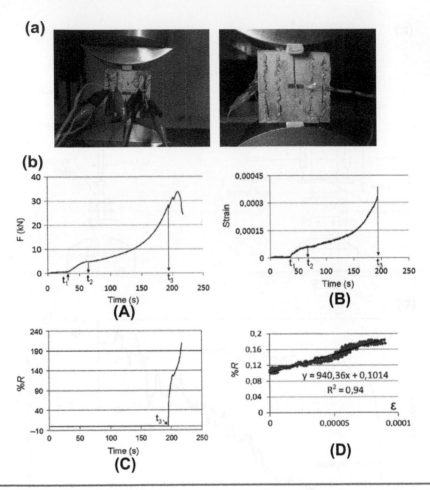

Figure 5.32   Sensing properties of concrete with steel fiber under split tension. (a) Experimental setup. (b) Test results: (A) force-time graph, (B) strain-time graph, (C) %R-time graph, and (D) the percent change of electrical resistance (%R–tensile strain graph) [37].

They pointed out that the increase in change of electrical resistance during loading is due to crack opening, whereas the decrease in change of electrical resistance during unloading is attributed to crack closure. The irreversibly increasing part of electrical resistance is due to the fiber/matrix interface weakening, whereas the reversible part is due to crack opening (as shown in Figure 5.35) [42].

Wen and Chung tested the variation of the fractional change in electrical resistivity with strain and stress for cement paste containing 0.36 vol.% and 0.72 vol.% steel fiber under repeated tension. Both resistivity and strain increase with the stress, and the results are partial reversible. The higher the stress amplitude, the higher the strain and the resistivity. However, the correlation between resistivity and strain

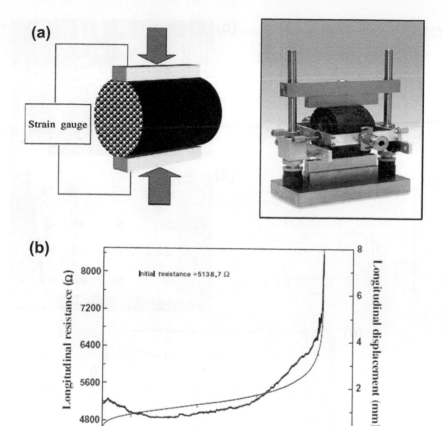

Figure 5.33   Sensing properties of asphalt concrete with graphite powder and carbon fiber under indirect tension. (a) Experimental setup [38,39]. (b) Test results [39].

depends on the load history. This dependence is undesirable for practical strain sensing. The resistivity and the strain are more reversible for the cement paste containing 0.36 vol.% steel fiber [43]. Zhou and Yang investigated the sensing behavior of carbon-coated nylon fiber concrete under cyclic uniaxial tension with different stress amplitudes. The change in electrical resistivity is reversible and stable for elastic deformation, and irreversible for non elastic deformation and fracture (as shown in Figure 5.36) [34].

Hou and Lynch tested the sensing property of the engineered cementitious composites with hybrid polyvinyl alcohol fiber and steel fiber under cyclic tension by using the AC method. They observed that the change in electrical resistivity is strongly correlated with the mechanical strain in a linear fashion. When the

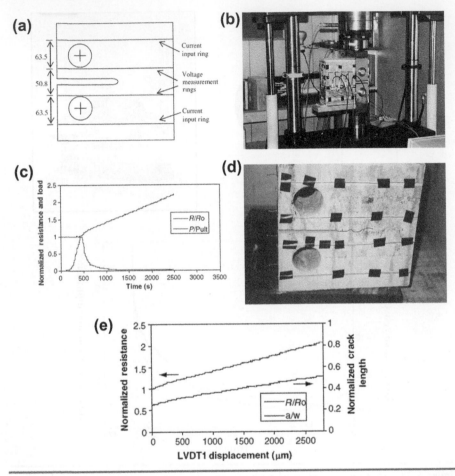

Figure 5.34   Sensing properties of concrete with carbon fiber under compact tension. (a) Specimen for compact tension. (b) Experimental setup. (c) Resistance and load vs time for specimen with 0.6% by volume of fibers tested at 10 kHz.(d) Test results: tortuous crack growth. (e) Normalized resistance and crack length vs displacement for specimen with fiber at 1 MHz [40].

composites are completely unloaded, the electrical resistivity of the composites is almost fully recovered (as shown in Figure 5.37) [35].

Saafi tested the sensing behavior of carbon nanotube cement paste under cyclic uniaxial tension. He observed that the electrical resistance first increased and then decreased, following the increase and decrease of the applied load. The composite exhibits good repeatability of sensing properties (as shown in Figure 5.38) [36].

Azhari observed the sensing property of concrete with carbon fiber. She obtained the same results as Saafi (as shown in Figure 5.39) [2].

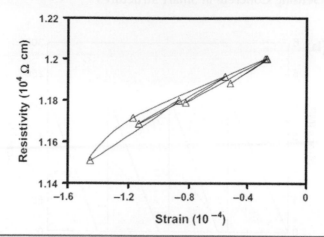

Figure 5.35 Sensing properties of concrete with carbon fiber under cyclic uniaxial tension [42].

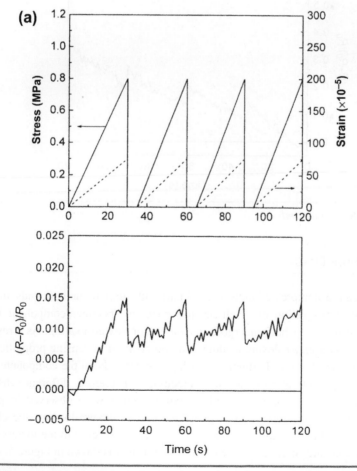

Figure 5.36 Sensing properties of concrete with carbon-coated nylon fiber under cyclic uniaxial tension. (a) Low-stress amplitude. (b) High-stress amplitude [34].

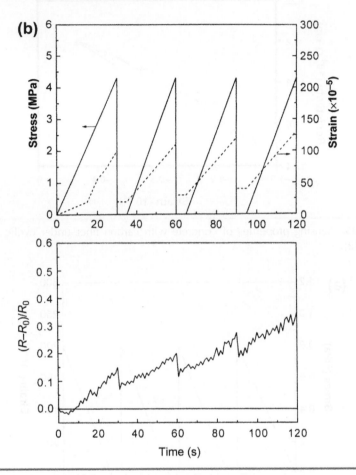

Figure 5.36   *(continued)*

5.2.3 UNDER FLEXURE

When a beam is compressed in the central part with supports on both ends, the upper half of the cross-section of a self-sensing concrete specimen/component is compressed and the lower half undergoes tension under flexure loading. As a result, the sensing behavior under flexure loading is a composition of sensing properties under compression and tension. Furthermore, it heavily depends on the components of the self-sensing concrete and the layout of electrodes. Therefore, different sensing behaviors of self-sensing concrete under flexure loading were observed in previous research. Wang et al. observed that under monotonic flexure loading, the electrical resistivity smoothly decreases as load increases, and it reaches to the lowest point at the ultimate flexural strength and then sharply turns up (as shown in Figure 5.40) [43].

(a)

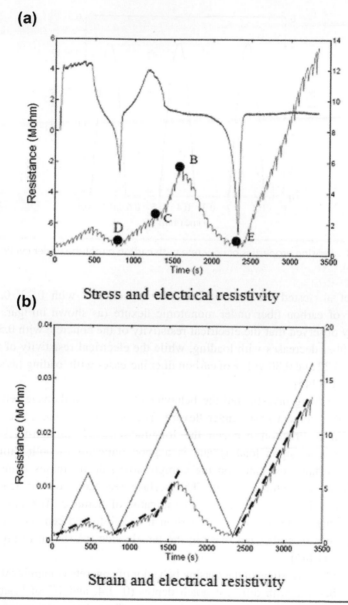

Stress and electrical resistivity

(b)

Strain and electrical resistivity

Figure 5.37 Sensing properties of concrete with carbon-coated nylon fiber under cyclic uniaxial tension. (a) Stress and electrical resistivity. (b) Strain and electrical resistivity [35].

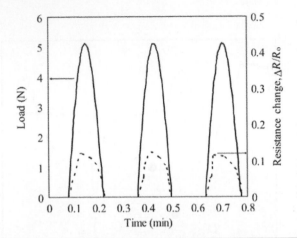

Figure 5.38 Sensing properties of concrete with carbon nanotube under cyclic uniaxial tension [36].

Chen et al. tested the sensing behavior of concrete with 0.22, 0.55, and 0.80 vol.% of carbon fiber under monotonic flexure (as shown inFigure 5.41(a) [44]). They observed that the electrical resistivity of the concrete with 0.55 vol.% of carbon fiber decreases with loading, while the electrical resistivity of the concretes with 0.22 and 0.80 vol.% of carbon fiber increases with loading (as shown in Figure 5.41(b)) [45].

Wen and Chung investigated the behavior of strain and damage self-sensing in carbon fiber cement paste under flexure. The oblique resistance (i.e., volume resistance in a direction between the longitudinal and through-thickness directions) increases upon loading and is a good indicator of both damage and strain. The surface resistance on the compressive side decreases upon loading and is a good indicator of strain. The surface resistance on the tensile side increases upon loading and is a good indicator of damage. The effectiveness for self-sensing of flexural strain in carbon fiber reinforced cement is enhanced by the presence of embedded rebar in the tensile side (as shown in Figure 5.42 and Table 5.1) [46].

Chen and Liu investigated the sensing behavior of concrete containing 0.80 wt.% of carbon fiber and with different notch depths (0, 1/4, and 1/2 of beam section height) in the tensile side under monotonic flexure. They observed that the electrical resistivity first decreases and then increases with increase of loading when the notch depth is 1/2 of beam section height. This is because the electrical resistivity is only the electrical resistivity in the compressive zone. As a result, the change of electrical resistivity under flexure acts like that under compression. The electrical resistivity of concrete without notch depth increases with loading. The strain in tensile zone is bigger than that in the compressive zone. The increase of electrical resistivity in the

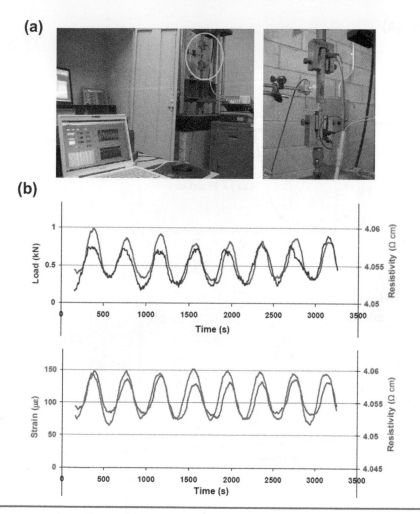

Figure 5.39 Sensing properties of concrete with carbon fiber under cyclic tension. (a) Experimental setup. (b) Test results [2].

tensile zone under loading dominates the change in electrical resistivity of the whole specimen. The electrical resistivity does not change much with the increase of loading when the notch depth is 1/4 of beam section height, which is because the resistance changes in the compressive zone and tensile zone affect and counteract each other (as shown in Figure 5.43) [47].

Chen and Liu also pointed that the relative change of resistance curves could be divided into three zones during the whole loading process: (1) the elastic stage of proportional increase of the deflection with the load. The resistance increases at a small range with the deflection increase. (2) the nonelastic stage with the occurrence

(a)

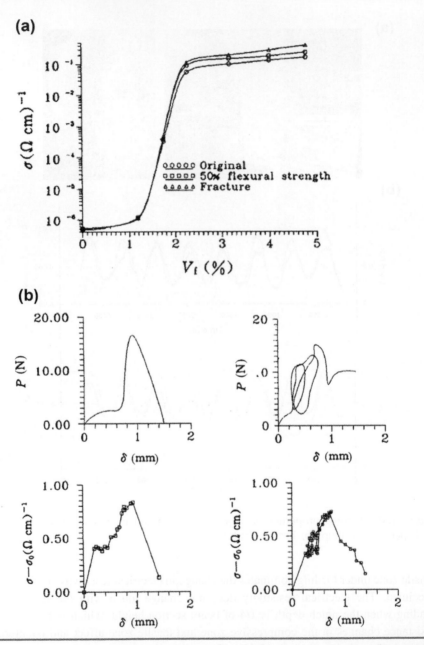

(b)

Figure 5.40 Sensing properties of concrete with carbon fiber under flexure. (a) Electrical conductivity vs fiber fraction under loading. (b) Load-relative changing of electrical conductivity: crossbeam displacement curves in three-point bending [43].

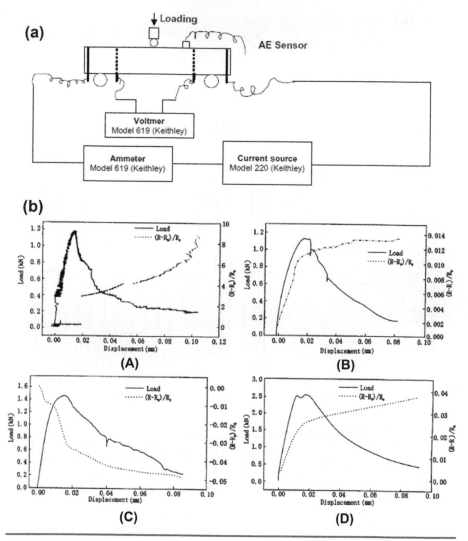

Figure 5.41   Sensing properties of concrete with carbon fiber under flexure. (a) Sketch map of experimental setup [44]. (b) Test results: (A) 0% of carbon fiber, (B) 0.20% of carbon fiber, (C) 0.55% of carbon fiber, and (D) 0.80% of carbon fiber [45].

of visible cracks. There is a turning point in both the load-deflection curves and the relative change of resistance curves. The load-deflection relation changes to be nonlinear, while the resistance rapidly increases with the deflection increase. (3) The descending stage of load-deflection curves. The resistance remains stable since the composite already fails and the inner conductive network is broken (as shown in Figure 5.44) [47].

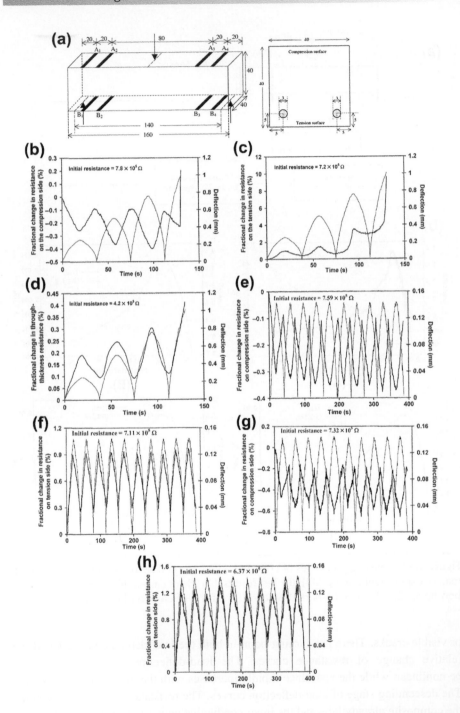

TABLE 5.1 Ratio (0.01/mm) of the Peak Value of the Fractional Change in Surface Resistance to the Midspan Deflection [46]

Midspan Deflection (mm)	Surface	Without Steel	With Steel
0.143	Tension	6.5 ± 0.8	9.3 ± 1.2
0.143	Compression	-2.4 ± 0.8	-4.1 ± 0.5
0.215	Tension	6.9 ± 0.8	9.6 ± 1.1
0.215	Compression	-2.6 ± 0.3	-4.3 ± 0.5

Azhari tested the sensing property of the concrete with 5 vol.% of carbon fiber subjected to flexure. It was observed that as the point load is applied and hence the displacement increases, there initially is a sharp decrease in the resistivity; then as the displacement values reach 0.3 mm, the decrease in resistivity becomes slower and eventually stops decreasing at 0.35 mm of vertical displacement. After that the resistivity values start to gradually increase with an increase in load and displacement until the specimen fails (as shown in Figure 5.45) [2].

The above research results indicate that there are good corresponding relationships between the electrical resistivity/resistance of the self-sensing concrete and its stress/strain, crack, damage, and damage extent. Therefore, the stress/strain, crack, and damage inside self-sensing concrete can be in situ monitored by measuring its electrical signals.

Figure 5.42 Sensing behavior of concrete with carbon fiber under flexure. (a) Sketch map of experimental setup. (b) Fractional change in surface resistance (thick curve) at the compression side vs time and deflection (thin curve) vs time during flexural loading at progressively increasing deflection amplitudes up to failure. (c) Fractional change in surface resistance (thick curve) at the tension side vs time and deflection (thin curve) vs time during flexural loading at progressively increasing deflection amplitudes up to failure. (d) Fractional change in through-thickness resistance (thick curve) vs time and deflection (thin curve) vs time during flexural loading at progressively increasing deflection amplitudes up to failure. (e) Variation of the fractional change in resistance on the compression side (thick curve) with time and of the deflection (thin curve) with time during repeated flexural loading for carbon fiber concrete with embedded steel. (f) Variation of the fractional change in resistance on the tension side (thick curve) with time and of the deflection (thin curve) with time during repeated flexural loading for carbon fiber concrete without embedded steel. (g) Variation of the fractional change in resistance on the compression side (thick curve) with time and of the deflection (thin curve) with time during repeated flexural loading for carbon fiber concrete with embedded steel. (h) Variation of the fractional change in resistance on the tension side (thick curve) with time and of the deflection (thin curve) with time during repeated flexural loading for carbon fiber concrete with embedded steel [46].

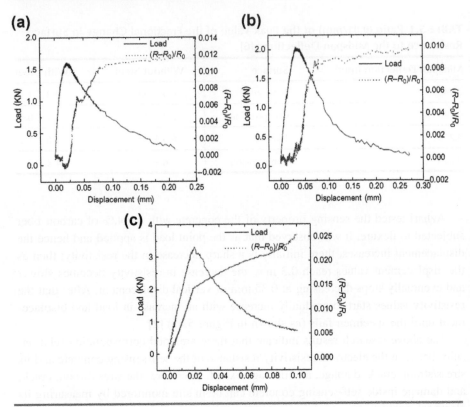

Figure 5.43 Sensing properties of concrete with carbon fiber under flexure. (a) Without notch. (b) Notch depth is 1/4 of beam section height. (c) Notch depth is 1/2 of beam section height [47].

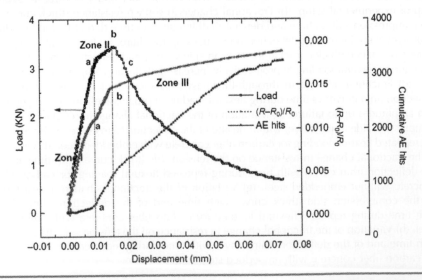

Figure 5.44 Sensing properties of concrete with carbon fiber under flexure [47].

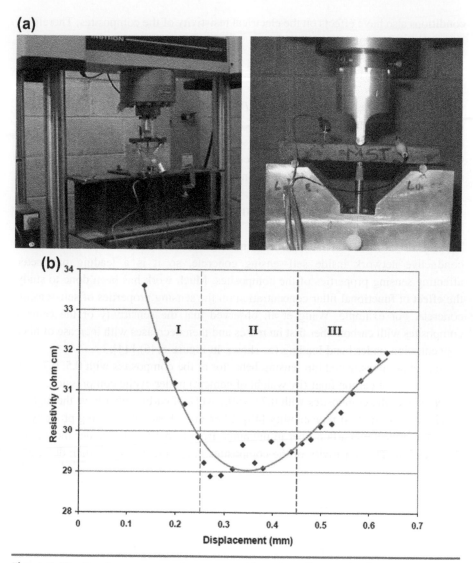

Figure 5.45 Sensing property of concrete with carbon fiber under flexure. (a) Experimental setup. (b) Test results [2].

5.3 Some Factors Affecting Sensing Properties

As shown in Eq. (5.1), external forces would lead to changes in the electrical resistivity of self-sensing concrete. However, besides external forces, some other factors such as composite components, processing technology, and environmental conditions also have effects on the electrical resistivity of the composites. Therefore, Eq. (5.1) can be generalized as [22]:

$$\frac{\Delta\rho}{\rho_0} = f(X, C, P, H, \cdots) \tag{5.2}$$

where C is the composite components, P represents processing technology, and H refers to environmental conditions.

Because some factors that influence sensing properties have been mentioned in Chapters 1 to 4, this section will introduce the detailed effect of the main factors on the sensing properties of self-sensing concrete.

5.3.1 FUNCTIONAL FILLER CONCENTRATION

The functional filler concentration dominates the formation and distribution of the conductive network inside self-sensing concrete, so it is a leading parameter affecting sensing properties of the composites. Much work has been done to study the effect of functional filler concentration on the sensing properties of self-sensing concrete. For example, Wang et al. observed that the sensitivity of the cement composites with carbon fiber first increases and then decreases with increase of fiber concentration under bend loading (as shown in Figure 5.46) [44].

Jian et al. investigated the sensing behavior of the composites with 0.5, 0.7, 0.9, and 1.1 wt.% of carbon fiber (by weight of cement) under cyclic compression. They observed that the composites with 0.7 and 0.9 wt.% of carbon fiber have the highest sensitivity and the best repeatability [48]. Chen et al. tested the sensing property of cement mortar with carbon fiber in the range from 0 to 1.6 vol.% under monotonic compression. The sensitivity of the composites firstly increases and then decreases

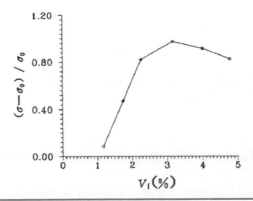

Figure 5.46 Relative changing of electrical conductivity with fiber volume fraction under loading [44].

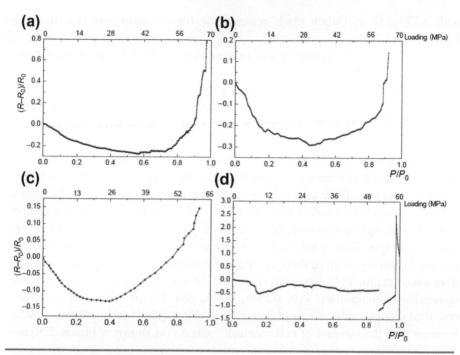

Figure 5.47 Relationship between fractional change in electrical resistance for carbon fiber concrete with different carbon fiber contents. (a) 0.2%. (b) 0.4%. (c) 0.55%. (d) 0.8% [23].

with an increase of fiber concentration. The stress level at which the $\max|\Delta\rho/\rho_0|$ occurs decreases with an increase of fiber concentration (as shown in Figure 5.47 and Table 5.2) [23].

Li et al. investigated the sensing properties of cement paste composites with 3.11, 6.04, 7.22, 8.79, 11.39, and 13.85 vol.% of carbon black. They observed that the composite with 3.11 vol.% of carbon black has no sensing property, while that

TABLE 5.2 Stress Level at which the Lowest $\Delta R/R$ Occurs [23]

Properties	Fiber Volume Fraction (%)		
	0.20	0.55	0.80
Stress level at which the lowest $\Delta R/R$ occurs (%)	40	45	78
Lowest $\Delta R/R$	−0.128	−0.290	−0.260
$\Delta R/R$ at carbon fiber concrete failure	0.146	0.192	0.796

with 8.79 vol.% of carbon black presents the highest sensitivity (as shown in Figure 5.48) [49].

Jia studied the sensing property of cement mortar with different concentrations of magnetic fly ash and steel slag under compression. The $\max|\Delta\rho/\rho_0|$ and sensitivity of this type of composite increase with concentration of magnetic fly ash and steel slag when the concentration of magnetic fly ash and steel slag is in the range from 0 to 80 wt.% and from 0 to 500 wt.%, respectively (as shown in Figure 5.49) [8].

Han et al. observed that the composite with 22 vol.% of nickel powder shows the highest sensitivity among the composites containing 20, 22, and 24 vol.% of nickel powder [50]. Deng tested the sensing behavior of cement paste and concrete with a concentration of carbon fiber from 0.8 to 1.2 wt.% (by cement weight) and of concrete with hybrid iron containing conductive functional aggregate (32 vol.%) and carbon fiber from 0.8 to 1.2 wt.% (by cement weight). He found that the sensing sensitivity of all of these composites decreases with an increase of carbon fiber concentration [51]. Lin et al. compared the sensing properties of engineered cementitious composites with 0.25%, 0.5%, and 1% of carbon black. They observed that generally, the gage factors in elastic segment and inelastic segment increase with the increase of carbon black contents (as shown in Figure 5.50 and Table 5.3) [36].

Jian et al. tested the sensing property of concrete with different concentrations of carbon fiber (0.5%, 0.7%, 0.9%, and 1.1%) under cyclic compression. They observed that the sensing sensitivity, repeatability, and stability are heavily dependent upon the carbon fiber concentrations. The concrete with 0.7% of carbon fiber had the best sensing properties [48].

Han et al. study the effect of concentration of carbon nanotube on the sensing property of composites. They observed that the change in amplitudes of the resistance for the three types of composites increases with the concentration of carbon nanotube (as shown in Figure 5.52) [7].

Teomete and Kocyigit compared the sensing properties of cement-based composites with different steel fiber concentrations under split tension. Experimental results indicate that there is a strong correlation between tensile strain and electrical resistance change (%R) (as shown in Figure 5.53). In general, the gage factor increases with fiber concentration. The highest gage factor of composite under tensile strain obtained is 5195. This is 2600 times higher than the gage factor of commercial metal strain gages, which have a gage factor of 2 (as shown in Figure 5.54(a)). The linearity has a general trend of decreasing by fiber concentration (as shown in Figure 5.54(b)). The correlation coefficients of best-fit line to the strain–%R graph are close to 1, which shows the strong linear relation between strain and %R. There is a general trend of increase of R^2 with the fiber concentration (as shown in Figure 5.54(c)) [37].

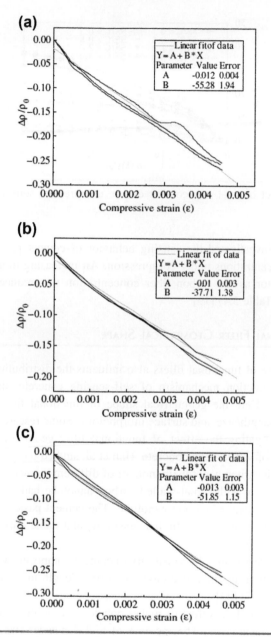

Figure 5.48 Relative change in electrical conductivity with fiber volume fraction under loading. (a) 15% of carbon black. (b) 20% of carbon black. (c) 25% of carbon black [49].

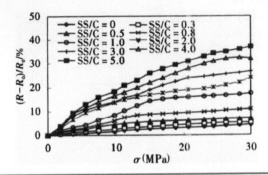

Figure 5.49 Effect of steel slag content on the compression sensitivity of steel slag concrete [8].

Baeza et al. investigated the sensing behavior of cement paste with different carbon fiber concentrations under compression. An increasing trend was observed for the gage factor as the carbon fiber concentration was reduced (as shown in Figure 5.55 and Table 5.4) [19].

5.3.2 FUNCTIONAL FILLER GEOMETRICAL SHAPE

Geometrical shape of functional fillers also influents the distribution of conductive network and conduction mechanism of self-sensing concrete, thus affecting its sensing behavior. Here the geometrical shape of functional fillers includes their shape/structure, length/size, and surface morphology. Some research effort has been devoted to investigating the effect of functional filler geometrical shape on the sensing property of self-sensing concrete. Han et al. studied the sensing behavior of cement paste with carbon nanotube/nanofiber of different diameters. They observed that the cement paste with signal-walled carbon nanotube had a higher sensitivity than that with multi-walled carbon nanotube. The cement paste containing carbon nanofiber of smaller diameter had higher sensitivity and stable sensing properties (as shown in Figure 5.56) [7].

Wang et al. observed that the sensitivity of cement composites with carbon fibers of 2, 4, 6, and 8 mm in length decreases with fiber length under bend loading (as shown in Figure 5.57) [44].

Han et al. explored the effect of nickel powder size on the sensing property of nickel powder cement paste. The increase of nickel powder size is beneficial for enhancing sensing sensitivity [50]. Jia compared the sensing property of cement mortar with grounding steel slag and original state steel slag. A decrease in steel slag size has little contribution to improvement of the sensitivity (as shown in Figure 5.58) [8].

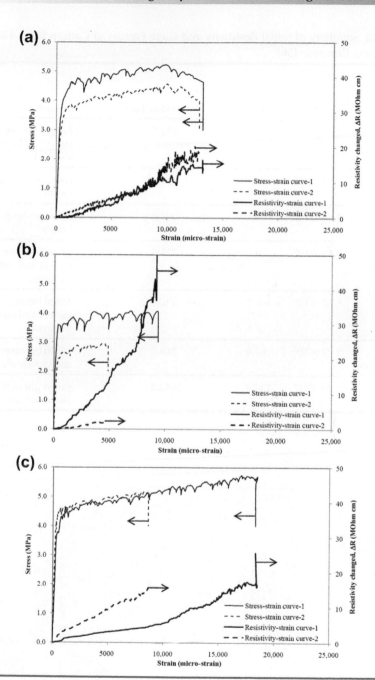

Figure 5.50   Relative change in electrical conductivity with fiber volume fraction under loading. (a) 0.25% of carbon black. (b) 0.5% of carbon black. (c) 1% of carbon black [36].

TABLE 5.3 Summary of Initial Resistivity and Gage Factors of Carbon Black-Filled Self-Sensing Concrete [36]

Specimen Type	Initial Resistivity (Ω cm)	Elastic Segment (Gage Factor)	Inelastic Segment (Gage Factor)	Remarks (As Shown in Figure 5.51)
Without carbon black	1.87×10^7	32.2	282.7	Failed between current and voltage electrode
With 0.25% carbon black	6.57×10^6	8.7	181.9	Failed within gage length
	4.84×10^6	198.1	284.3	Failed within gage length
With 0.5% carbon black	5.65×10^6	39.6	725.5	Failed between current and voltage electrode
	3.60×10^6	101.3	140.0	Failed between current and voltage electrode
With 1.0% carbon black	2.27×10^6	67.8	706.2	Failed between current and voltage electrode
	1.29×10^6	237.2	692.2	Failed just outside aluminum plate

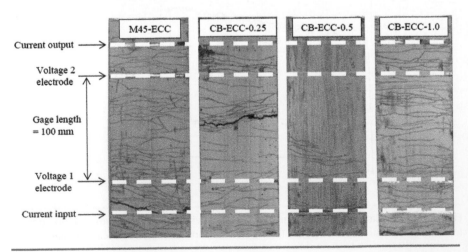

Figure 5.51 Cracking pattern of specimens [36].

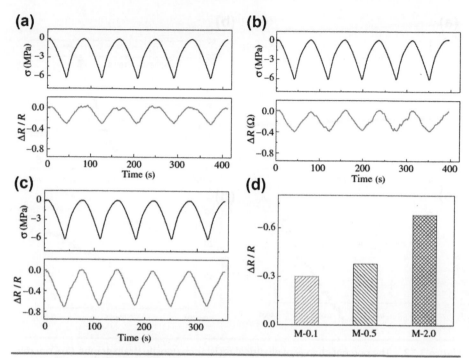

Figure 5.52 Sensing property of concrete with different concentrations of carbon nanotube. (a) 0.1% (M-0.1). (b) 0.5% (M-0.1). (c) 2.0% (M-0.1). (d) Amplitude of resistance change [7].

Han et al. investigated the sensing behavior of cement paste with normal smooth spherical and spiky spherical nickel powders. They found that the cement paste with smooth spherical nickel powders has no, or weak, sensing properties, while that with spiky spherical nickel powders possesses strong sensing properties. This is because the sharp nano-tip on the surface of spiky spherical nickel particles can induce field emission and tunneling effects, which lead to the highly sensitive responses to compressive stress and strain [52].

Gao et al. compared the sensing properties between concrete with unoxidized carbon fiber (CNT PR-19-XT-PS), oxidized carbon (CNT PR-19-XT-PS-OX), and heat-treated carbon fiber (CNT PR-19-XT-LHT-OX) (as shown in Figure 5.59 and Table 5.5). CNT PR-19-XT-PS is produced by pyrolytically stripping the fiber as produced to remove polyaromatic hydrocarbons from the fiber surface. CNT PR-19-XT-PS-OX is an oxidized version of CNT PR-19-XT-PS, and is more conductive than CNT PR-19-XT-PS. CNT PR-19-XT-LHT-OX is produced by heat-treating the fiber at 1500 °C. Heat treatment converts any chemical vapor-deposited carbon present on the surface of the fiber to a short-range ordered structure. The inherent conductivity of the fiber is increased following heat treatment, and it is the most

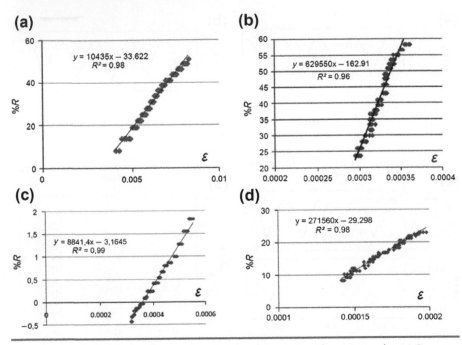

Figure 5.53 The percent change of electrical resistance–tensile strain graphs. (a) Concrete with steel fiber length L = 6 mm, 0.5% volume fraction. (b) Concrete with steel fiber length L = 6 mm, 0.8% volume fraction. (c) Concrete with steel fiber length L = 6 mm, 1.0% volume fraction. (d) Concrete with steel fiber length L = 6 mm, 1.5% volume fraction [37].

conductive of the three types of fibers. They observed that for SCC containing the CNF PR-19-XT-PS, resistance did not change much at CNF concentrations of 0.5% and 1.0%, which means the fiber concentration is still too low. In this case, the minimum concentration of CNF is 1.5%; otherwise, the electrical resistance variation cannot be detected. For SCC containing the CNF PR-19-XTPS-OX, comparatively steady electrical resistance variation can be detected when the concentration is between 1% and 2%; the minimum electrical resistance variation across this range of concentrations varies between 0.2 and 0.25 (Ω/Ω). Resistance variation decreases for all concentrations of fiber, and the minimum electrical resistance variations are inversely related to fiber concentration. The electrical resistance variation does not change much when the fiber concentration varies between 1% and 2%, but decreases significantly when the fiber concentration is higher than 2%. For SCC containing the CNF PR-19-XT-LHT-OX, resistance decreases for all concentrations of fiber. At 1.0% fiber concentration, CNFSCC10-LO exhibits the highest electrical sensitivity, with an 80% reduction in electrical resistance. Electrical resistance variation decreases rapidly when the fiber concentration is increased from 1.0% to 1.5%. This

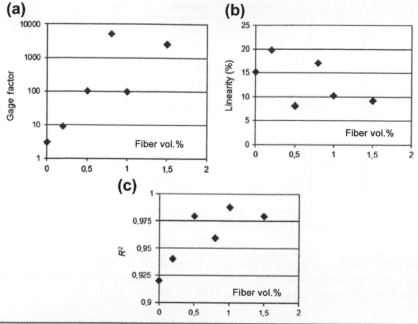

Figure 5.54 Relationship between sensing property and fiber concentration. (a) Gage factor. (b) Linearity (it is the percent of maximum difference between the input–output curve (%R vs strain curve) and fitted linear regression line, to full-scale output). (c) Correlation coefficient (R^2) [37].

trend continues at a slower rate for fiber concentrations greater than 1.5%. When the concentration of PR-19-XT-LHT-OX fibers exceeds 2.0%, the percent reduction in electrical resistance is less than 10% [11].

Baeza et al. tested the sensing property of cement paste containing carbon fiber with 3, 6, and 12 mm length under compression. As the carbon fiber concentration is 0.5%, the composite with 3 mm of carbon fiber (i.e., the lowest fiber aspect) has the highest gage factor among composites with three kinds of fiber length (as shown in Table 5.4) [19].

5.3.3 LOADING RATE

A higher loading rate would limit crack propagation and inhibit plastic deformation of cementitious composites. This may alter the change trend of the conductive network inside self-sensing concrete under loading, thus influencing sensing behavior of the concrete. Cao and Chung tested sensing property of carbon fiber cement mortar at loading rates of 0.144, 0.216, and 0.575 MPa/s. They observed that an increase in strain rate causes the fractional change in electrical resistivity at any

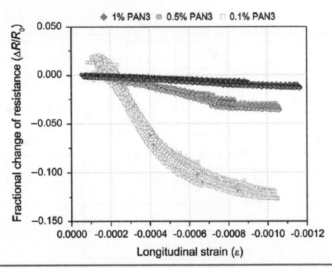

Figure 5.55 Resistance fractional change vs longitudinal strain for cement pastes rein-forced with 3 mm long carbon fiber at different concentrations: 0.1%, 0.5%, and 1.0% carbon fiber with respect to cement mass [19].

strain level to decrease, in addition to causing the fractional change in electrical resistivity at failure to decrease (as shown in Figure 5.60 and Table 5.6) [53].

Han et al. studied the effect of three loading rates (0.2, 0.4, and 0.6 mm/min) on the sensing property of cement-based composite with nickel powder. The sensing property of this composite was found to be almost free from the effect of the loading rate (as shown in Figure 5.61) [54].

Jia et al. investigated the sensing behavior of steel slag concrete under different loading rates including 0.1, 0.5, 1.0, and 2.0 kN/s. They observed that the $\max|\Delta\rho/\rho_0|$, $[(\rho-\rho_0)/\rho_0]/\sigma$, and linearity of the sensing property all increase with loading rates. The bigger the loading rate, the higher the compressive stress corresponding to $\max|\Delta\rho/\rho_0|$. The balance state is more obvious at a lower loading rate, while there is no balance state (i.e., an abrupt increase state directly follows the previous decrease state) at a higher loading rate (as shown in Figure 5.62) [29].

Han et al. observed the sensing behavior of the carbon nanotube cement-based composite at loading rates of 0.05, 0.10, 0.15, 0.20, 0.25, and 0.30 cm/min under repeated compressive loadings with amplitude of 4 MPa. The sensing behavior is almost free from the effect of loading rate when the loading rate is lower than 0.20 cm/min. If the loading rate exceeds 0.20 cm/min, it will affect the self-sensing responses of the composite, and this effect increases with the loading rate [55]. Azhari and Banthia tested the sensing property of the concrete with carbon fiber under loading rates of 0.01, 0.02, 0.04, 0.06, and 0.12 kN/s. They observed that the

TABLE 5.4 Gage Factor of Cement Pastes Under Different Compression Stress, Fiber Length, and Concentration with Respect to Cement Mass [19].

Concentration of Carbon Fiber (%)	Fiber Length (mm)	Stress (MPa)	Gage Factor	
			Loading Cycles	Unloading Cycles
1.0	3	1.25	7.1	7.5
1.0	3	2.50	9.0	9.4
1.0	3	5.00	11.5	11.3
1.0	3	7.50	10.2	9.7
1.0	3	10.00	11.3	10.7
0.5	3	1.25	26.1	23.8
0.5	3	2.50	39.7	40.8
0.5	3	5.00	40.7	40.6
0.5	3	7.50	42.7	45.4
0.5	3	10.00	41.3	40.6
0.1	3	1.25	105.8	116.2
0.1	3	2.50	152.8	173.0
0.1	3	5.00	158.6	178.1
0.1	3	7.50	148.8	163.8
0.1	3	10.00	139.6	144.4
0.5	6	1.25	4.3	4.1
0.5	6	2.50	11.0	12.3
0.5	6	5.00	16.4	17.5
0.5	6	7.50	16.4	16.9
0.5	6	10.00	16.5	16.6
0.5	12	1.25	14.4	16.2
0.5	12	2.50	24.4	24.8
0.5	12	5.00	29.5	30.8
0.5	12	7.50	29.8	31.0
0.5	12	10.00	28.8	30.2

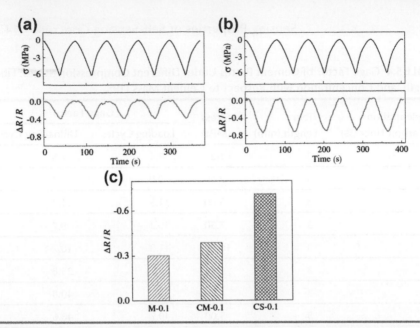

Figure 5.56 Sensing property of concrete with different types of carbon nanotube. (a) 0.1% of carboxyl multiwall carbon nanotube (CM-0.1). (b) 0.1% of carboxyl single-wall carbon nanofiber (CS-0.1). (c) Amplitude of resistance change (M-0.1 means 0.1% of multi-wall carbon nanotube) [7].

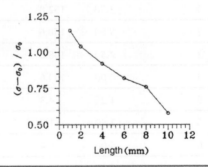

Figure 5.57 Relative changing of electrical conductivity with fiber length under loading [44].

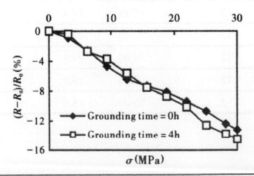

Figure 5.58 Effect of steel slag grounding time (which affects the slag size) on the compression sensitivity of concrete with steel slag [8].

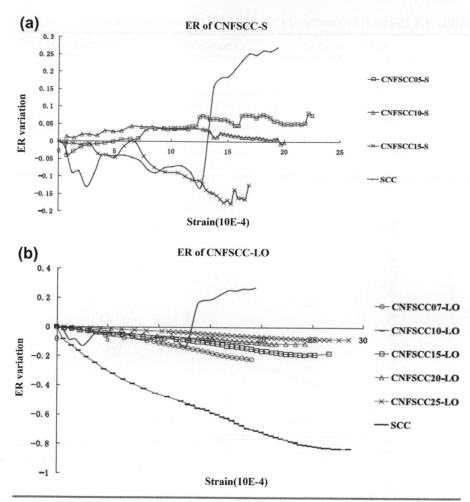

Figure 5.59 Sensing properties of concrete with different types of carbon fiber. (a) Concrete with unoxidized carbon nanofiber. ER is electrical resistance; SCC denotes self-consolidating concrete. CNFSCC05-S, CNFSCC10-S, and CNFSCC15-S denote self-consolidating concrete containing the CNF PR-19-XT-PS in amounts of 0.5%, 1.0%, and 1.5% by volume of binder, respectively). (b) Concrete with heat-treating carbon nanofiber. SCC denotes self-consolidating concrete. CNFSCC07-LO, CNFSCC 10-LO, CNFSCC15-LO, CNFSCC20-LO, and CNFSCC25-LO denote self-consolidating concrete containing the CNF PR-19-XT-LHT-OX in amounts of 0.7%, 1.0%, 1.5%, 2.0%, and 2.5% by volume of binder, respectively [11].

sensing property of this composite is also loading rate-dependent (as shown in Figure 5.63) [5].

Baeza et al. studied the relationship between loading rate and gage factor of three cement pastes with different carbon lengths (0.5% PAN12, 0.5% PAN6, and 1.0%

TABLE 5.5 Electrical Properties of Concrete with Carbon Nanofiber [11]

	SCC	CNFSCC-S				CNFSCC-PO				CNDSCC-LO				
Concentration of CNF %	0	0.25	0.5	1.0	1.5	1.0	1.5	2.0	2.5	0.7	1.0	1.5	2.0	2.5
Average-R (kΩ)	–	–	–	8.03	5.66	1.92	0.95	0.57	0.52	1.67	1.91	0.92	0.95	0.51
Resistivity (k$\Omega \cdot$m)	–	–	–	0.84	0.59	0.20	0.10	0.06	0.05	0.18	0.20	0.10	0.10	0.05
$\Delta R/R$ (%)	–	–	–	0.22	0.33	0.22	0.22	0.26	0.11	0.24	0.85	0.13	0.11	0.10

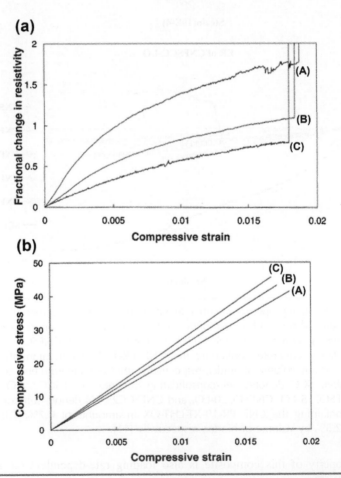

Figure 5.60 Carbon fiber concrete during compressive testing up to failure at loading rates of (A) 0.144, (B) 0.216, and (C) 0.575 MPa/s. (a) Fractional change in resistivity vs strain, (b) Stress vs strain [53].

TABLE 5.6 Effect of Strain Rate on Compressive Properties [53]

Loading Rate (MPa/s)	Strain Rate $(10^{-5}/s)$	Strength (MPa)	Modulus of Elasticity (GPa)	Ductility (%)	Fractional Change in Resistivity at Fracture
0.144	5.3	41.4 ± 1.6	1.83 ± 0.17	1.9 ± 0.2	1.78 ± 0.24
0.216	8.8	43.2 ± 1.0	1.85 ± 0.14	1.8 ± 0.2	1.10 ± 0.13
0.575	23.3	45.7 ± 2.1	1.93 ± 0.17	1.8 ± 0.3	0.81 ± 0.16

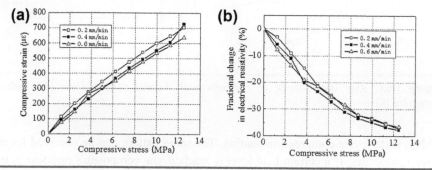

Figure 5.61 Relationship between stress/strain and fractional change in electrical resistivity of concrete with nickel powder under different loading speeds. (a) Stress–strain. (b) Resistivity–stress [54].

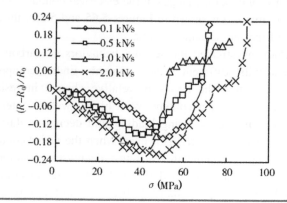

Figure 5.62 Effect of loading rate on the compression sensitivity of concrete with steel slag [29].

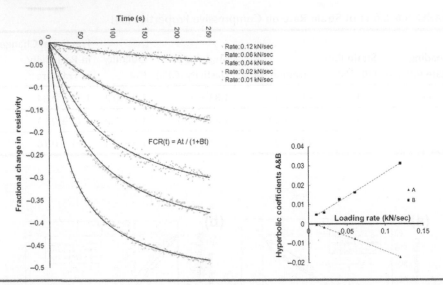

Figure 5.63 Rate dependence of concrete with hybrid carbon fiber and carbon nanotube [5].

PAN3) under cycle compression loading. The same pattern can be observed for all three composites: if a certain load level is reached, the composite's response is independent of the loading rate used (as shown in Figure 5.64) [19].

5.3.4 WATER CONTENT

The water content inside the self-sensing concrete depends on many factors such as environmental humidity and temperature, curing regime, and structures of the concrete. Its variation would cause changes in the electrical conductivity of functional fillers and concrete matrix (as shown in Figures 5.65 and 5.66), thus changing the sensing property of the self-sensing concrete.

Wang and Zhao investigated the sensing behaviors of carbon fiber cement-based composites with different water contents under compression. Their research results show that the trends of relative change in resistance of the composites vary with the water content. With higher water content, the relative change in resistance increases during loading and decreases during unloading, showing irregular sensing property. However, when the water content is reduced after drying, the relative change in resistance decreases monotonically during loading, and increases monotonically during unloading, demonstrating the regular sensing property [15]. Li [58] and Jia [29] observed that the sensing properties of concrete with steel slag, magnetic fly ash, and hybrid steel slag and magnetic fly ash, under water saturation condition, surface-dry condition, air- dried condition,

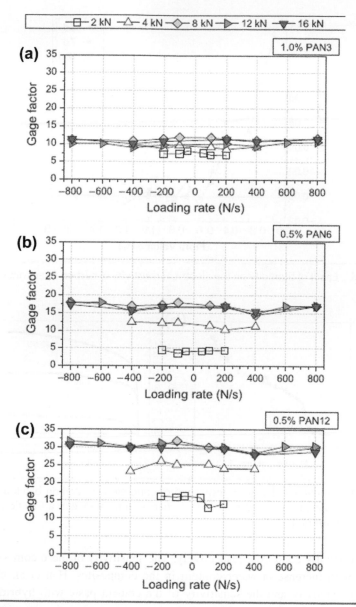

Figure 5.64 Mean value of gage factor of carbon fiber cement pastes vs loading rate (negative values correspond to unloading cycles) for different maximum axial loads (the stress levels applied were 1.25, 2.5, 5.0, 7.5, and 10.0 MPa for 2, 4, 8, 12, and 16 kN loads, respectively). (a)With 1% of 3 mm carbon fiber (PAN 3). (b) With 0.5% of 6 mm carbon fiber (PAN6). (c) With 0.5% of 12 mm carbon fiber (PAN12) [19].

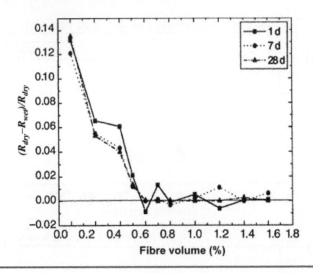

Figure 5.65 Effect of humidity on the electrical resistivity of carbon fiber concrete [56].

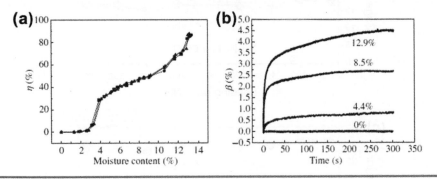

Figure 5.66 Fractional change in resistance of concrete with carbon black as a function of water contents (a), relative fractional change in resistance as a function of measurement time (b) [57].

and absolutely dry condition, respectively. The sensitivity of these composites all increases with increase in water content in the composites. Han et al. observed that the conductivity and the sensitivity of the cement paste with hybrid carbon fiber and carbon black are enhanced as the water content in the composites increases [59].

In addition, Han et al. compared the responses of electrical resistance of the MWNT/cement composites with different water contents (0.1, 1.3, 3.3, 5.7, 7.6, and 9.9%) to compressive stress under repeated compressive loadings with amplitude of 6 MPa. The piezoresistive sensitivities of these MWNT/cement composites first increase, and then decrease, with the increase of water content in the composites [60].

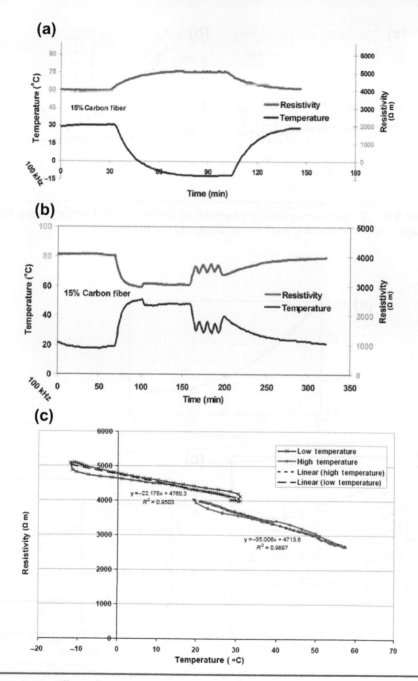

Figure 5.67 Effect of temperature variation on the electrical resistivity of carbon fiber concrete. (a) Low temperature. (b) High temperature. (c) Reversibility of the temperature effect [2].

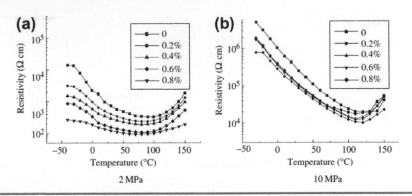

Figure 5.68 Temperature–resistivity properties of concrete with different carbon fiber contents under compressing pressure. (a) 2MPa. (b) 10MPa [61].

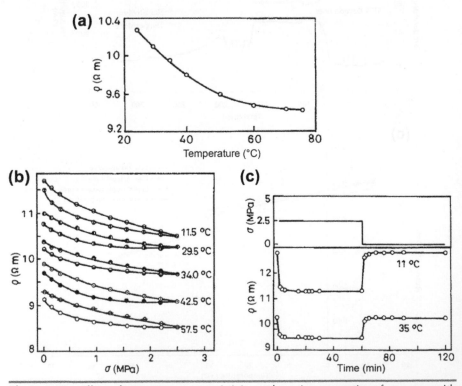

Figure 5.69 Effect of temperature on resistivity and sensing properties of concrete with carbon fiber. (a) Relation between resistivity and temperature. (b) Relation between resistivity and stress. (c) Resistivity curves by loading and unloading suddenly under a constant stress at different temperatures [62].

5.3.5 TEMPERATURE

A rise or decrease in temperature can result in expansion or contraction of the self-sensing concrete, which would alter distance between adjacent functional fillers. Moreover, a change in temperature will also cause an increase or decrease in the transition energy of the electrons in functional fillers, and alter the water content of the concrete matrix. Therefore, the electrical conductivity (as shown in Figures 5.67 and 5.68) [2,61] and sensing properties of self-sensing concrete are heavily concerned with temperature. Mao et al. tested the electrical conductivity and sensing properties of carbon fiber cement paste at different temperatures (11.5, 29.5, 34.0, 42.5, and 57.5 °C). They observed that the electrical resistivity and sensitivity of this composite decreases with temperature (as shown in Figure 5.69) [62]. Jia studied the sensing behavior of the concrete with only steel slag and hybrid steel slag and magnetic fly ash at temperatures of −20, 5, 25, and 50 °C. He found that this concrete has higher sensing sensitivity at high temperature [29].

5.3.6 OTHER FACTORS AFFECTING SENSING PROPERTIES

Besides the factors above mentioned, Mao et al. tested the sensing properties of concrete with carbon fiber under a constant stress during the period of loading and unloading [4].

Li [51] and Jia [29] investigated the effect of loading duration at different stress amplitudes (as shown in Figure 5.70), size of tested specimen, spacing between electrodes, freeze–thaw cycle, dry–wet cycle, and corrosive environment on the sensing properties of concrete with steel slag.

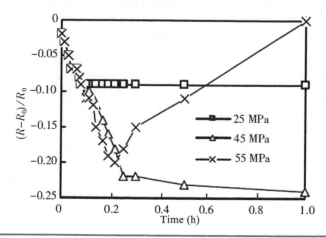

Figure 5.70   Effect of loading duration on sensing properties of steel slag concrete [29].

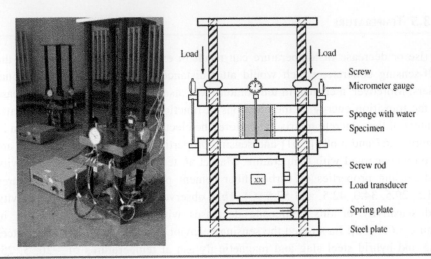

Figure 5.71 Schematic of long-term loading experimental setup.

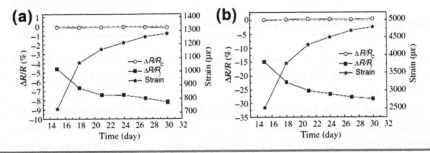

Figure 5.72 Fractional change in resistance ($\Delta R/R$, where R is resistance) and creep behavior of carbon black concrete encapsulated with epoxy during long-term loading with a stress ratio of 0.2 (a), and with a stress ratio of 0.66 (b) in a moist environment. (R_i = initial resistance; R_C = compensated resistance) [57].

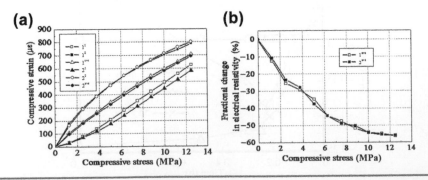

Figure 5.73 Relationship between stress and strain and between strain and fractional change in electrical resistivity of concrete with nickel powder under eccentric compression. (a) Stress–strain (b) Resistivity–stress. (1^1, 1^2, and 1^{ave} represent strains measured by strain gages 1, 2, and their average value, respectively, at first loading; and 2^1, 2^2, and 2^{ave} represent strains measured by strain gages 1, 2, and their average value, respectively, at second loading.) [54].

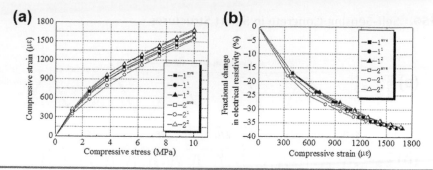

Figure 5.74 Relationship between stress and strain and between strain and fractional change in electrical resistivity of concrete with nickel powder under eccentric compression. (a) Stress–strain. (b) Resistivity–strain. (1^1, 1^2, and 1^{ave} represent strains measured by strain gages 1, 2, and their average value, respectively, at first loading; and $2^1, 2^2$, and 2^{ave} represent strains measured by strain gages 1, 2, and their average value, respectively, at second loading.) [63].

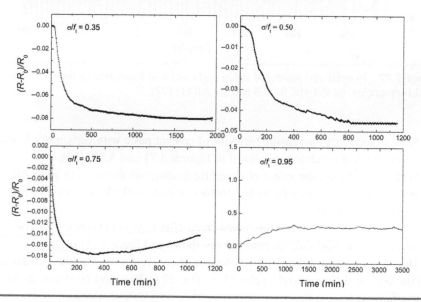

Figure 5.75 Resistivity creep behavior of carbon fiber concrete with fiber volume fraction 0.55% under various uniaxial pressures (σ and f_t are stress and ultimate stress, respectively) [21].

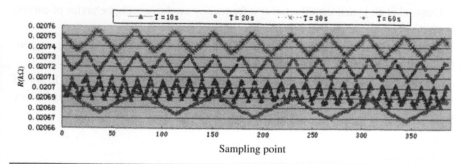

Figure 5.76 Sensing characteristics of concrete with graphite powder and carbon fiber under different load cycles [24].

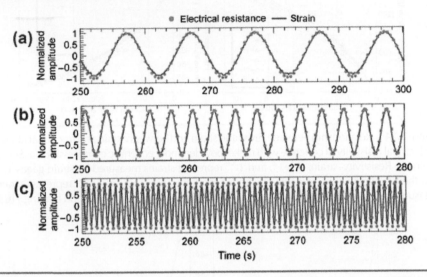

Figure 5.77 Normalized measured strain and electrical resistance vs time for different load frequencies. (a) 0.1 Hz. (b) 0.5 Hz. (c) 2.0 Hz [27].

Li et al. explored the sensing behavior of cement paste with carbon black under long-term loading condition (as shown in Figures 5.71 and 5.72) [57].

Han et al. studied the effect of eccentric loading on the sensing properties of cement-based composites with nickel powder or carbon black (as shown in Figures 5.73 and 5.74) [54,63].

Chen et al. observed that a resistance creep effect exists in carbon fiber concrete in the case of the load holding constant (as shown in Figure 5.75) [21].

Huang et al. studied the sensing behavior of concrete with graphite powder and carbon fiber at different load cycles (T = 10 s, 20 s, 30 s, and 60 s) (as shown in Figure 5.76) [24].

Materazzi et al. investigated the electrical response of carbon nanotube cement paste subjected to sinusoidal axial compression load in the load frequency range 0.1–5 Hz (as shown in Figure 5.77) [27].

Demirel et al. tested the effect of test frequency on the sensing behavior of carbon fiber concrete (as shown in Figure 5.78) [64].

Han et al. investigated the effect of test current on the sensing property of concrete with carbon black (as shown in Figure 5.79) [63].

Baeza et al. studied the effect of compression stress level on the gage factor of carbon fiber cement paste (as shown in Figure 5.64), and the effect of loading history on the sensing property (as shown in Figure 5.80) [19].

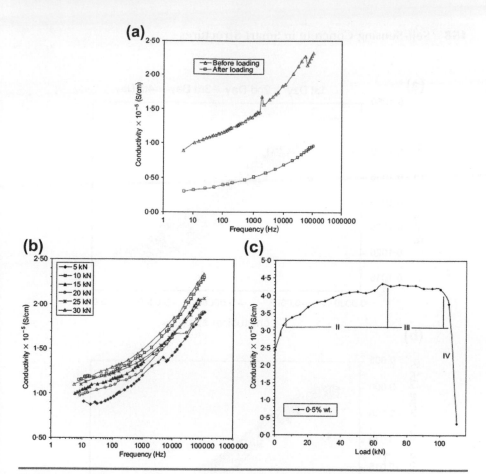

Figure 5.78 Electrical behavior of carbon fiber concrete with increasing loading in varying and constant frequencies under compression (a) Frequency vs conductivity of concrete with carbon fiber curves before and after loading (at the compressive load up to one-third of the compressive strength). (b) Change in conductivity with increasing frequency during the loading. (c) Change in conductivity with increasing loading at the constant frequency (40 kHz) [64].

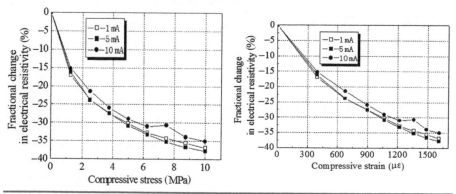

Figure 5.79 Relationship between stress/strain and fractional change in electrical resistivity of concrete with carbon black [63].

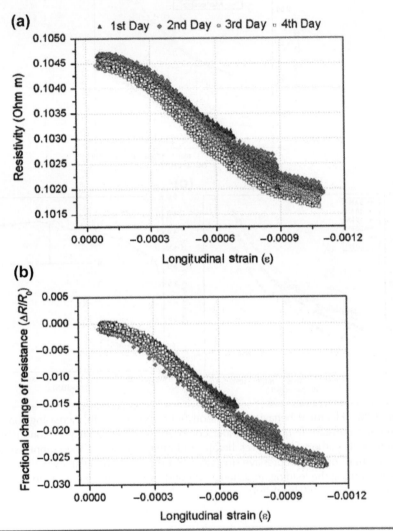

Figure 5.80 Resistivity vs longitudinal strain (a) and resistance fractional change vs strain (b) for carbon cement paste on different consecutive days [19].

▌ 5.4 Summary and Conclusions

The sensing properties of self-sensing concrete can be described by the response of electrical properties to external force or deformation. This response can be evaluated through some parameters such as sensitivity input/output range, linearity, repeatability, hysteresis, signal-to-noise ratio, and zero shifts. The sensitivity is the most important parameter to characterize the sensing property of self-sensing concrete. The gage factor of self-sensing concrete can reach hundreds or even thousands.

The sensing behavior of self-sensing concrete depends on the loading condition. Generally, the electrical resistance decreases under compression and increases under tension. The sensing behavior under flexure loading is a composition of sensing properties under compression and tension. In addition, the sensing property is affected by some factors such as functional filler concentration, functional filler geometrical shape, loading rate, water content, temperature, etc.

There are good corresponding relationships between the electrical signals and load/deformation of self-sensing concrete under different temporal and spatial conditions. Strain, stress, cracks, and damage inside the self-sensing concrete can therefore be in situ monitored by measuring electrical signals of the self-sensing concrete.

References

[1] Han BG, Ou JP. Embedded piezoresistive cement-based stress/strain sensor. Sens Actuators A 2007;138(2):294–8.

[2] Azhari F. Cement-based sensors for structural health monitoring. Dissertation for the master degree of applied science. Canada: University of British Columbia; 2008.

[3] Chung DDL. Self-monitoring structural materials. Mater Sci Eng Reports 1998;22(2):57–78.

[4] Mao QZ, Zhao BY, Sheng DR, Li ZQ. Resistance changement of compression sensible cement speciment under different stresses. J Wuhan Univ Technol 1996;11:41–5.

[5] Azhari F, Banthia N. Cement-based sensors with carbon fibers and carbon nanotubes for piezoresistive sensing. Cem Concr Compos 2012;34:866–73.

[6] Li GY, Wang PM, Zhao XH. Pressure-sensitive properties and microstructure of carbon nanotube reinforced cement composites. Cem Concr Compos 2007;29:377–82.

[7] Han BG, Zhang K, Yu X, Kwon E, Ou JP. Fabrication of piezoresistive CNT/CNF cementitious composites with superplasticizer as dispersant. J Mater Civ Eng 2012;24(6):658–65.

[8] Jia XW, Qian JS, Tang ZQ. Research on compression sensitivity of steel slag concrete. Mater Rev 2008;22(11):122–4.

[9] Cheng X, Wang SD, Huang SF, Chen W. Influence of forming technology on piezoresistivity of CFSC. J Build Mater 2008;11(1):84–8.

[10] Li H, Xiao HG, Ou JP. A study on mechanical and pressure-sensitive properties of cement mortar with nanophase materials. Cem Concr Res 2004;34(3):435–8.

[11] Gao D, Sturm M, Mo YL. Electrical resistance of carbon-nanofiber concrete. Smart Mater Struct 2009;18:095039.

[12] Han BG, Han BZ, Ou JP. Novel piezoresistive composite with high sensitivity to stress/strain. Mater Sci Technol 2010;26(7):865–70.

[13] Chen PW, Chung DDL. Carbon fiber reinforced concrete as a smart material capable of nondestructive flaw detection. Smart Mater Struct 1993;2:22–30.

[14] Hong L. Study on the smart properties of the graphite slurry infiltrated steel fiber concrete. Dissertation for the doctor degree in engineering. China: Dalian University of Technology; 2006.

[15] Wang YL, Zhao XH. Positive and negative pressure sensitivities of carbon fiber-reinforced cement-matrix composites and their mechanism. Acta Mater Compos Sin 2005;22(4):40–6.

[16] Wu B, Huang XJ, Lu JZ. Biaxial compression in carbon-fiber-reinforced mortar, sensed by electrical resistance measurement. Cem Concr Res 2005;35(7):1430–4.

[17] Zheng LX, Song XH, Li ZQ. Study on the compression sensibility of CFRC under quasi-triaxial compression. Bull Chin Ceram Soc 2004;4:40–3.

[18] Jia XW, Tang ZQ, Qian JS. Investigation on the strain sensitivity of steel slag concrete under axial load. J Funct Mater 2008;39(8):1344–7.

[19] Baeza FJ, Galao O, Zornoza E, Garcés P. Effect of aspect ratio on strain sensing capacity of carbon fiber reinforced cement composites. Mater Des 2013;51:1085–94.

[20] Chung DDL. Damage in cement-based materials, studied by electrical resistance measurement. Mater Sci Eng 2003;42:1–40.

[21] Chen B, Wu KR, Yao W. Piezoresistivity in carbon fiber reinforced cement based composites. J Mater Sci Technol 2004;20(6):746–50.

[22] Xiao HG, Li H, Ou JP. Self-monitoring properties of concrete columns with embedded cement-based strain sensors. J Intell Mater Syst Struct 2011;22(2):191–200.

[23] Chen B, Liu JY, Wu KR. Electrical responses of carbon fiber reinforced cementitious composites to monotonic and cyclic loading. Cem Concr Res 2005;35:2183–91.

[24] Huang L, Gong XJ, Sun MQ. Experimenttal research on the piezoresistive properties of cement-based composites with different conductive admixtures. Bull Chin Ceram Soc 2009;28(6):1112–7.

[25] Han BG, Yu X, Kwon E, Ou JP. Piezoresistive Multi-Walled Carbon Nanotubes filled cement-based composites. Sens Lett 2010;8(2):344–8.

[26] Materazzi AL, Ubertini F, D'Alessandro A. Carbon nanotube cement-based transducers for dynamic sensing of strain. Cem Concr Compos 2013;37:2–11.

[27] Wang LN, X. W, Hui GC, Li LQ. Research on compression sensitivity of carbon fiber reinforced concrete in triaxial cyclic loading. Concrete 2011;257(3):14–6.

[28] Jia XW, Qian JS, Tang ZQ. Research and mechanism analysis on the compression sensitivity of steel slag concrete. Mater Sci Technol 2010;18(1):66–70.

[29] Jia XW. Electrical conductivity and smart properties of Fe1-σO waste mortar. Dissertation for the doctor degree in engineering. China: Chongqing University; 2009.

[30] Meehan DG, Wang SK, Chung DDL. Electrical-resistance-based sensing of impact damage in carbon fiber reinforced cement-based materials. J Intell Mater Syst Struct 2010;21(1):83–105.

[31] Chen PW, Chung DDL. Concrete as a new strain/stress sensor. Composites:Part B 1996;27B:11–3.

[32] Sett K. Characterization and modeling of structural and self-monitoring behavior of fiber reinforced polymer concrete. Dissertation for the master of science in civil engineering. USA: University of Houston; 2003.

[33] Zhou ZJ, Xiao ZG, Pan W, Xie ZP, Luo XX, Jin L. Carbon-coated-nylon-fiber-reinforced cement composites as an intrinsically smart concrete for damage assessment during dynamic loading. J Mater Sci Technol 2003;19(6):583–6.

[34] Hou TC, Lynch JP. Conductivity-based strain monitoring and damage characterization of fiber reinforced cementitious structural components. Proc SPIE 2005;5765:419–29.

[35] Saafi M. Wireless and embedded carbon nanotube networks for damage detection in concrete structures. Nanotechnology 2009;20:395502 (7pp).

[36] Lin VWJ, Mo L, Lynch JP, Li VC. Mechanical and electrical characterization of self-sensing carbon black ECC. Proc SPIE – Int Soc Opt Eng 2011;7983:798316 (12pp).

[37] Teomete E, Kocyigit OI. Tensile strain sensitivity of steel fiber reinforced cement matrix composites tested by split tensile test. Constr Build Mater 2013;47:962–8.

[38] Liu XM, Wu SP, Ye QS, Qiu J, Li B. Properties evaluation of asphalt-based composites with graphite and mine powders. Constr Build Mater 2008;22:121–6.

[39] Liu XM, Wu SP. Study on the graphite and carbon fiber modified asphalt concrete. Constr Build Mater 2011;25:1807–11.

[40] Farhad R, Jerry AY, Gordon BB. Electrical resistance change in compact tension specimens of carbon fiber cement composites. Cem Concr Compos 2004;26:873–81.

[41] Chen PW, Chung DDL. Carbon fiber reinforced concrete as an intrinsically smart concrete for damage assessment during dynamic loading. J Am Ceram Soc 1995;78(3):816–8.

[42] Wen SH, Chung DDL. Piezoresistivity-based strain sensing in carbon fiber reinforced cement. ACI Mater J 2007;104(2):171–9.

[43] Wen SH, Chung DDL. A comparative study of steel- and carbon-fibre cement as piezoresistive strain sensors. Adv Cem Res 2003;15(3):119–28.

[44] Wang XF, Wang YL, Jin ZH. Electrical conductivity characterization and variation of carbon fiber reinforced cement composite. J Mater Sci 2002;37:223–7.

[45] Chen B, Liu JY. Damage in carbon fiber-reinforced concrete, monitored by both electrical resistance measurement and acoustic emission analysis. Constr Build Mater 2008;22:2196–201.

[46] Chen B, Wu KR, Yao W. Studies on the elelctrical conductivity of carbon fiber reinforced concrete and its applications. Concrete 2002;153(7):23–5.

[47] Wen SH, Chung DDL. Self-sensing of flexural damage and strain in carbon fiber reinforced cement and effect of embedded steel reinforcing bars. Carbon 2006;44(8):1496–502.

[48] Jian HL, Xie HC, Liu JW. The effect of fiber content on the mechanical properties and pressure-sensitivity of CFRC. Concrete 2003;170(12):21–2.

[49] Li H, Xiao HG, Ou JP. Effect of compressive strain on resistance of carbon black filled cement-based composites. Cem Concr Compos 2006;28:824–8.

[50] Han BG, Han BZ, Yu X. Effects of content level and particle size of nickel powder on the pie-zoresistivity of cement-based composites/sensors. Smart Mater Struct 2010;19:065012 (6pp).

[51] Deng X. Preparation and performance investigation of iron containing aggregate and its cement-based conductive composites. Dissertation for the master degree in engineering. China: Wuhan University of Technology; 2011.

[52] Han BG, Han BZ, Yu X, Ou JP. Ultrahigh pressure-sensitive effect induced by field emission at sharp nano-tips on the surface of spiky spherical nickel powders. Sens Lett 2011;9(5):1629–35.

[53] Cao JY, Chung DDL. Effect of strain rate on cement mortar under compression, studied by electrical resistivity measurement. Cem Concr Res 2002;32(5):817–9.

[54] Han BG, Lin Z, Ou JP. Piezoresistivity of cement-based materials with nickel powder. Rare metal materials and engineering. Rare Metal Mater Eng 2009;38:265–70.

[55] Han BG, Yu X, Zhang K, Kwon E, Ou JP. Sensing properties of CNT filled cement-based stress sensors. J Civ Struct Health Monit 2011;1:17–24.

[56] Chen B, Wu KR, Yao W. Conductivity of carbon fiber reinforced cement-based composites. Cem Concr Compos 2004;26:291–7.

[57] Li H, Xiao HG, Ou JP. Electrical property of cement-based composites filled with carbon black under long-term wet and loading condition. Compos Sci Technol 2008;68:2114–9.

[58] Li CT. Study on conductivity and strain sensitivity of steel-slag concrete. Dissertation for the master degree in engineering. China: Chongqing University; 2004.

[59] Han BG, Zhang LY, Ou JP. Influence of water content on conductivity and piezoresistivity of cement-based material with both carbon fiber and carbon black. J Wuhan Univ Technol-Mater Sci Ed 2010;25(1):147–51.

[60] Han BG, Yu X, Ou JP. Effect of water content on the piezoresistivity of CNTs/cement composites. J Mater Sci 2010;45:3714–9.

[61] Huang SF, Xu DY, Chang J, Xu RH, Lu LC, Cheng X. Smart properties of carbon fiber reinforced cement-based composites. J Compos Mater 2007;41(1):125–31.

[62] Mao QZ, Chen PH, Zhao BY, Li ZQ, Shen DR. Compresssion-sensitivity and temperature-sensitivity of carbon fiber reinforced cement under low stresses. Chin J Mater Res 1997;11(3):322–4.

[63] Han BG, Chen W, Ou JP. Study on piezoresistivity of cement-based materials with acetylene carbon black. Acta Mater Compos Sin 2008;25(3):39–44.

[64] Demirel B, Yazicioğlu S, Orhan N. Electrical behavior of carbon fibre-reinforced concrete with increasing loading in varying and constant frequencies. Mag Concr Res 2006;58(10):691–7.

Sensing Mechanisms of Self-Sensing Concrete

Chapter Outline

6.1 Introduction and Synopsis

The sensing mechanism is the foundation of understanding and controlling the self-sensing property of concrete. Since the concept of self-sensing concrete was proposed, a large amount of work has been done to explore the generation mechanism of the sensing property. The sensing property results from the change of the **163** conductive network inside the concrete composite under external force or deformation. The mechanism of electrical conduction and sensing property for concrete-based composites are similar to those of other conductive composites such as polymer-based or ceramic-based composites. However, there are some differences in the conductive characteristics of matrix materials, distribution characteristics of fillers in the matrix, and bonding characteristics between matrix and fillers for the composites. The conductive characteristics of concrete materials depend on their compositions and structures. The distribution characteristics of fillers in the concrete matrix depend on the compositions and processing technology of the composites. The bonding characteristics between the concrete matrix and fillers depend on the compositions of the composites, the surface conditions of fillers, and processing technology of the composites. This chapter will introduce the mechanism of electrical conduction and sensing properties of self-sensing concrete from four aspects: electrical conduction, conductive mechanism without loading and under external force, and constitutive model of sensing characteristic behavior of self-sensing concrete.

6.2 Type of Electrical Conduction

The basic types of electrical conduction of self-sensing concrete include electronic and/or hole conduction (i.e., contacting conduction, tunneling conduction, and/or field emission conduction) and ionic conduction. Electrons and/or holes come from conductive functional fillers, while ions come from the cement concrete matrix. In addition, it should be noted that hole conduction is only effective for carbon fillers such as carbon fiber, carbon nanotube/nanofiber, and carbon black [1–4].

6.2.1 CONTACTING CONDUCTION

This type of conduction is due to the direct contact of neighboring functional fillers, thus forming a conductive link. It is associated with the motion of electrons and/or holes through the conductive paths formed by functional fillers that are tiny and contact each other. The microstructural observation of self-sensing concrete has provided direct evidence for the existence of contacting conduction. Contacting conduction has been widely used in explaining the conductive behavior of self-sensing concrete with different functional fillers [5–7].

6.2.2 TUNNELING CONDUCTION AND FIELD EMISSION CONDUCTION

Tunneling conduction takes place when electrons jump through the energy barriers between functional fillers in concrete matrix [1,8,9]. Some researchers theorize that field emission is a manifestation of the tunneling effect [10,11]. However, because field emission is induced by a local strong electric field, other researchers consider that field emission is different from quantum tunneling to some extent [12–14]. Tunneling conduction and field emission conduction both are associated with the transmission conduction of electrons between the disconnected yet close-enough fillers. Tunneling conduction has contributed to the electrical conductivity of self-sensing concrete with different functional fillers [1,8,9], whereas field emission conduction is not widely used because conventional functional fillers cannot generate a strong electric field to induce field emission at applied low voltages. However, some functional fillers with unique morphology can induce a localized increase of the electric field at sharp tips, which effectively reduces the barrier's width and allows field emission conduction to occur. If the projections on the surface of the functional filers have tip radii below 10 nm, the local electric field at this sharp nano-tip can have a field enhancement factor as high as 1000. Li et al. [3] and Han et al. [15] considered that the field emission conduction from the nano-scale tip of the carbon nanotube has contributed to the electrical conductivity of self-sensing concrete with carbon nanotubes. Han et al. proved that the field emission is an important conduction mechanism for self-sensing concrete containing spiky, spherical nickel powder with sharp nano-tips on its surface [16]. Until now, many researchers have ascribed some conductive behaviors of self-sensing concrete to tunneling conduction and field emission conduction mechanisms theoretically, and some observed experimental results also support this theory well [1–3,15,16].

6.2.3 IONIC CONDUCTION

Hydrated cement paste, in addition to calcium silicate hydrate gel and other solid phases, contains a variety of voids. The water filling these voids or pores can dissolve ionic species (mainly Ca^{2+} and OH^-) from the solid phases, resulting in some ionic conduction through the interconnected capillary pores. Since the ionic conduction is associated with the motion of ions in pore solution, ionic conductivity varies in a particularly wide range when cement contains a substantial amount of free water. In dry conditions, the cement matrix approximates an insulating material [17,18]. Many researches have proved that electrical conduction in the intrinsic self-sensing cement concrete, especially in concrete with a filler concentration below the percolation threshold, involves ionic conduction [6,19]. The ionic conduction leads to electric polarization, which induces an increase of the measured DC electrical resistivity of self-sensing concrete during resistance measurement. The polarization strongly depends on the type and concentration of functional filler, and components of the concrete matrix. For example, Wen and Chung observed that crystalline carbon fiber increases the extent of polarization (as shown in Figure 6.1, the electrical resistivity changes with the time of polarization for different carbon fillers and matrixes) [20].

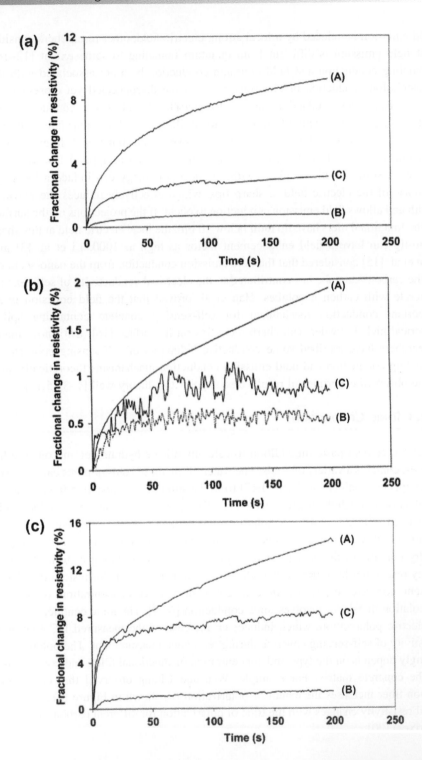

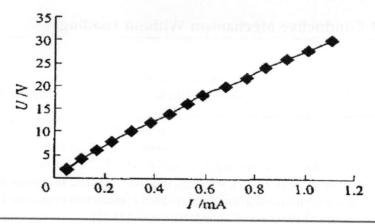

Figure 6.2 AC voltage–current characteristic of concrete with carbon fiber [21].

It should be noted that the actual electrical conduction mechanism of self-sensing concrete is very complex in nature. The above-motioned conduction types coexist in the composite and interrelate with each other. The DC electrical resistance–time relationship can indicate, between the electronic hole conduction and the ionic conduction, which one dominates in the electrical conductivity of self-sensing concrete. When ionic conduction is dominant, the DC electrical resistance increases obviously with measurement time due to the polarization effect; in the meantime, the AC electrical resistance is constant (as shown in Figure 6.2) [21]. When electronic hole conduction is dominant, the DC electrical resistance basically keeps stable with measurement time. In addition, the current–voltage relationship can give an indication of whether the electrical conductivity of self-sensing concrete is due to tunneling and field emission conduction or direct contact of neighboring functional fillers. A linear current–voltage relationship indicates that the direct contact of neighboring functional fillers is the dominant conduction mechanism. In contrast, tunneling and field emission would induce a nonlinear power law current–voltage relation in the electrical conductivity of self-sensing concrete.

Figure 6.1 Variation of measured resistivity with time (polarization) for different types of carbon fiber cement pastes. (a) For cement pastes with fiber (0.5% by weight of cement) and silica fume: (A) amorphous pristine carbon fiber, (B) crystalline pristine carbon fiber, and (C) crystalline intercalated carbon fiber. (b) For cement pastes with fiber (1.0% by weight of cement) and latex: (A) amorphous pristine carbon fiber, (B) crystalline pristine carbon fiber, and (C) crystalline intercalated carbon fiber. (c) For cement pastes with fiber (0.5% by weight of cement) and latex: (A) amorphous pristine carbon fiber, (B) crystalline pristine carbon fiber, and (C) crystalline intercalated carbon fiber [20].

■ 6.3 Conductive Mechanism Without Loading

Self-sensing concrete is fabricated by adding functional fillers into concrete matrix, and its conductive characteristics are closely related to the concentration of functional fillers. The change of electrical resistivity of the self-sensing concrete along with filler concentration, i.e., the conductive characteristic curve of self-sensing concrete, is graphically represented in Figure 6.3. The conductive characteristic curve describes the percolation phenomenon, which has been observed in self-sensing concrete with different functional fillers (as shown in Figures 6.4–6.8) [2,22,23]. This curve can be divided into three sections. Zone A, with high resistivity, is called an insulation zone. Zone B, with sharply decreasing resistivity, is called a percolation zone. Zone C, with stabilized low resistivity, is called a conductive zone [24,25].

Zone A: insulation zone, when filler concentration in the concrete matrix is much lower than the percolation threshold, spacing between functional fillers is large, and gathering of the fillers is limited, so the conductive path is hard to form; the electrons have difficulty moving between fillers, and the composite exhibits almost the same high resistivity as the concrete matrix. The electrical conductance of the concrete matrix dominates the electrical conductance of the composites. For a cement

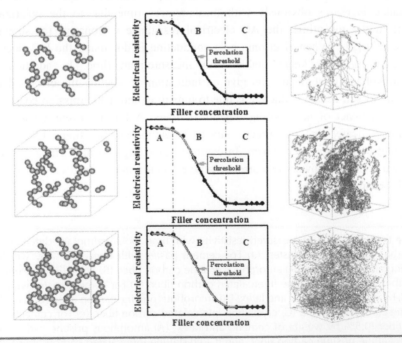

Figure 6.3   Change of the electrical resistivity along with filler concentration (the top row is in zone A, middle row is in zone B, and the bottom row is in zone C).

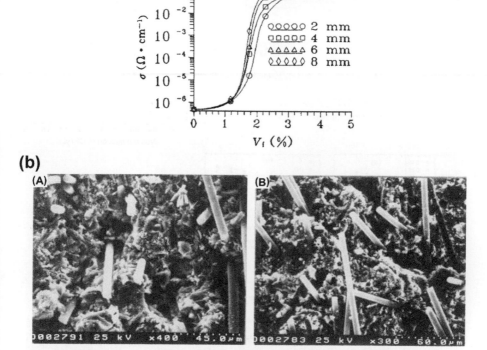

Figure 6.4 Change of the electrical conductivity of concrete with different lengths of carbon fiber along with filler concentration. (a) Conductive characteristic curves. (b) SEM photos: (A) 1.16% of carbon fiber, (B) 2.21% of carbon fiber. [22]

concrete matrix, the ionic conduction is dominant in the electrical conductivity of the composite in this zone.

Zone B: percolation zone, along with continuous increase of the filler concentration in the composite, spacing between adjacent fillers decreases and fillers start forming a conductive path. Probability of the electronic transition greatly rises, resulting in a sharply increase in conductivity of the composites. Contacting conduction, tunneling conduction and/or field emission conduction, and ionic conduction are all dominant factors in the electrical conductivity of the composite when the filler concentration is below the percolation threshold. However, when the filler concentration exceeds the percolation threshold, tunneling conduction and/or field emission conduction play leading roles in the composite conductivity in addition to the direct contact of functional fillers.

Zone C: conductive zone, when the filler concentration is much higher than the percolation threshold, functional fillers can be approximately regarded as totally contacting each other. The direct contact of functional fillers becomes the only dominant conduction mechanism.

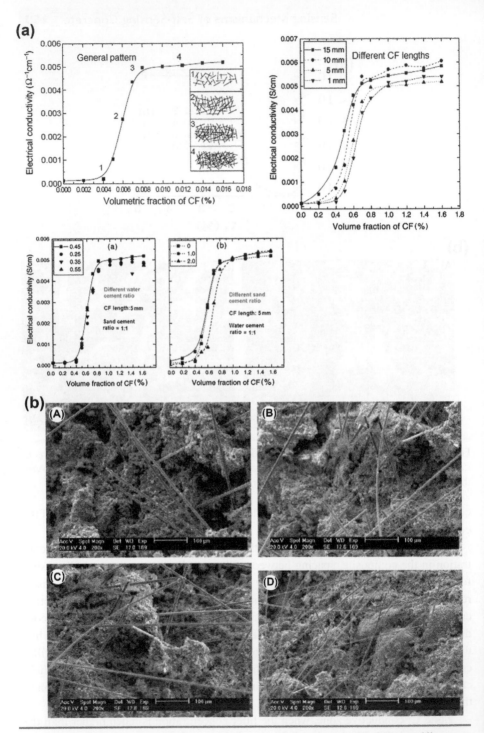

Figure 6.5 Change of the electrical conductivity of concrete with different filler concentration levels. (a) Conductive characteristic curves. (b) SEM photos: (A) 0.20% of carbon fiber, (B) 0.40% of carbon fiber, (C) 0.55% of carbon fiber, and (D) 0.80% of 5 mm carbon fiber [26].

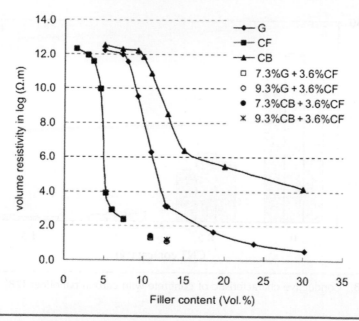

Figure 6.6 Conductive characteristic curves of asphalt concrete with different functional fillers (G, graphite powder; CF, carbon fiber; and CB, carbon black) [27].

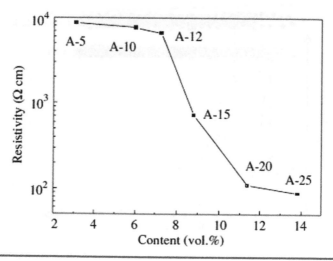

Figure 6.7 Conductive characteristic curves of concrete with carbon black [2].

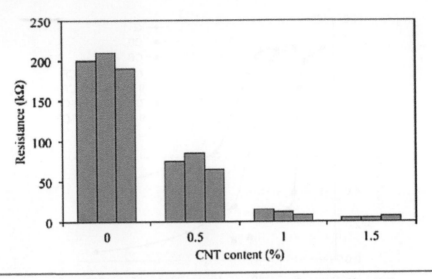

Figure 6.8 Conductive characteristic of concrete with carbon nanofiber [28].

It should be noted that the patterns of relationship curve between electrical resistivity and filler concentration have some difference for different functional fillers. Generally, fibrous fillers, having a high aspect ratio, can modify the electrical conductivity of concrete at a much lower concentration level compared with particle fillers (e.g., as shown in Figure 6.9) [29].

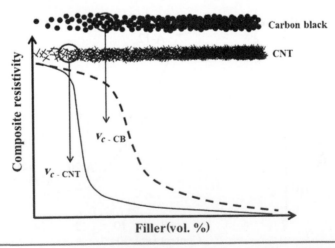

Figure 6.9 Illustration of carbon nanotube network percolation (v_c) compared with carbon black [29].

6.4 Conductive Mechanism under External Force

The electrical resistivity of self-sensing concrete changes when the concrete deforms under loading. Several factors may contribute to the change in electrical resistivity.

1. **Change of intrinsic resistance of functional fillers.** When an external force is applied to the composites, the fillers are deformed, resulting in a change of their intrinsic resistance. Jiang et al. observed that a single carbon fiber changes noticeably with strain under tension (as shown in Figure 6.10) [30]. Tombler et al. [31] and Pushparaj et al. [32] have reported that the deformation of carbon fibers can cause remarkable changes in their intrinsic electrical resistance (as shown in Figure 6.11).

   Li et al. also discussed that the change of intrinsic resistance of carbon nanotubes contributes to the sensing behavior of the composites [3]. However, Han et al. considered that the change of intrinsic resistance of nickel powder can be negligible because of the extremely small deformation of nickel powder compared to concrete matrix [33].

2. **Change of bonding between functional fillers and matrix.** When the composite is subjected to external force, the bonding between filler and matrix will alter, thus causing the contact resistance to change. Chung considered that the sensing property of self-sensing concrete with carbon fiber stems from this reason. The resistivity in both stress direction and transverse direction increases upon tension

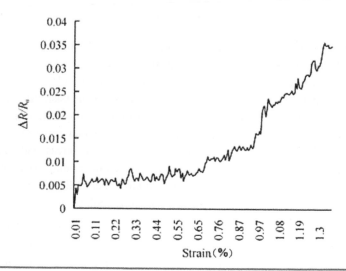

Figure 6.10   The change of electrical resistance of a single carbon fiber with strain [30].

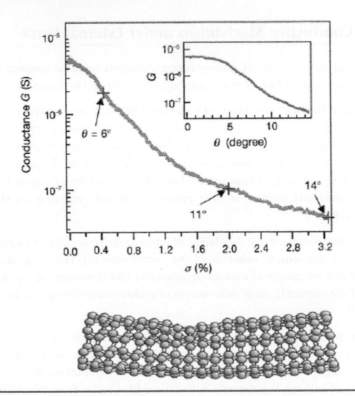

Figure 6.11 Conductance change induced by mechanical deformation of metallic carbon nanotube using AFM tip (G, conductance of carbon nanotube; σ, stain of carbon nanotube) [31].

because of slight fiber pull-out that accompanies crack opening, and decreases upon compression due to slight fiber push-in that accompanies crack closure [34]. As shown in Figure 6.12, Fu and Chung tested the relationships of shear stress vs. displacement (solid curve) and contact electrical resistivity vs. displacement (dashed curve) simultaneously obtained during pull-out testing of carbon fiber from cement paste. They observed that the contact resistivity gradually increased prior to the abrupt increase when the shear stress had reached its maximum. The stress also gradually increased as debonding took place, and reached its maximum when the fiber–matrix debonding was completed, i.e., the contact resistivity increased as debonding took place [35].

Fu and Chung further tested the variation of contact electrical resistivity with bond strength between fiber and concrete. They observed that the contact electrical resistivity is heavily dependent on the interface bonding strength between fiber and concrete matrix [36]. Jiang et al. also tested the relationship between the fractional

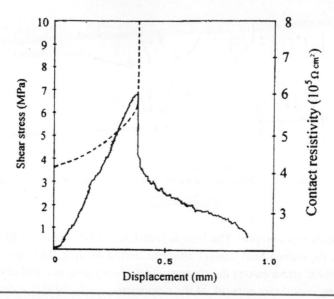

Figure 6.12   Shear stress vs. displacement (solid curve) and contact electrical resistivity vs. displacement (dashed curve) simultaneously obtained during pull-out testing of carbon fiber from cement paste [35].

increase in resistance and displacement during single carbon fiber pull-out testing [30]. As shown in Figure 6.13, they observed that the resistance increases with increasing interfacial bonding force. When the interfacial bonding force reaches its maximum, the bond between fiber and matrix is broken completely, and the

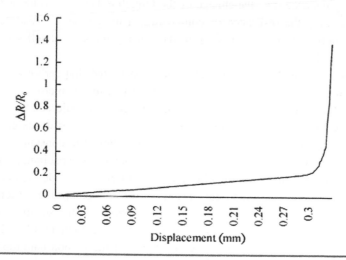

Figure 6.13   Relationship between the fractional increase in resistance and displacement during single fiber pull-out testing [30].

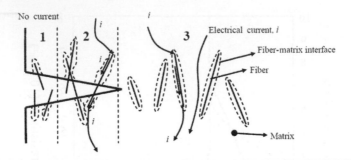

Figure 6.14 Fiber distribution in concrete matrix (2 denotes direct contact of the fibers) [37].

resistance increases abruptly. The load is transferred from matrix to fiber through interface in the carbon fiber cement concrete during loading. The variation of the interfacial shear stress causes the change of interfacial structure and results in the change of the conductive network of the composite.

3. **Change of contact between functional fillers**. The load will lead to the direct contact or separation of functional fillers (as shown in Figure 6.14), thus causing decrease or increase of the contact resistance. This factor has been considered as one of the main contributors to sensing behavior of self-sensing concrete.

4. **Change of tunneling distance between functional fillers**. Under applied external force, thickness and/or microstructures of the insulating concrete matrix between adjacent fillers may change considerably. This causes a change in the tunneling distance (i.e., the change of the tunneling barrier's width and height), thus changing the resistance of composites. This has been considered as the most important factor contributing to the sensing property of self-sensing concrete [1,6,11,16].

5. **Change of capacitance**. Wang and Zhao considered that carbon fiber can be regarded as the capacitance plates [6]. There is a lot of micro-capacitance in carbon fiber cement concrete due to the presence of ionic conduction in the cement concrete matrix. When the composite is subjected to external force, the capacitance plate distance and the relative dielectric constant of the concrete matrix will change, thus causing the change in capacitance (as shown in Figure 6.15). This would lead to a change of current inside the composites, and further alter the electrical resistance of the composite. Zheng et al. considered that the capacitance has a contribution to the sensing properties of the carbon fiber cement paste in the AC measurement [21]. However, Han et al. tested the response of capacitance of the carbon nanotube cement paste to compression, and found that capacitance is insensitive to compressive loading [38].

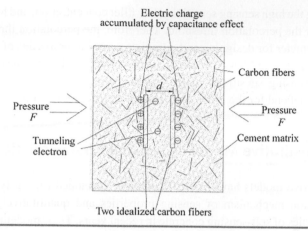

Figure 6.15   Micro-mechanism of double-effect mode [6].

In fact, the above-mentioned factors may work together toward the sensing properties of the self-sensing concrete, but only one or several of them are leading at different zones of the conductive characteristic curve.

Zone A: The conductive path is hard to form, even though an external force is applied to the composites. As a result, the composites present high initial electrical resistivity and possess no, or poor, sensing response to external force or deformation.

Zone B: At the beginning of zone B, the change of capacitance, the change of intrinsic resistance of fillers, and the change of bonding between filler and matrix are the dominant factors. Near the percolation threshold, the change of tunneling distance between fillers becomes the leading factor. At the end of zone B, the change of contact between fillers, the change of tunneling distance between fillers, and the change of intrinsic resistance of fillers become the leading factors. It can be seen from the above analysis that several factors together are in charge of the sensing properties of self-sensing concrete in each section of Zone B. Therefore, the composites at zone B have good sensing properties.

Zone C: The conductive fillers become crowded and are more likely to come into contact with each other. The change of contact between fillers and the change of intrinsic resistance of fillers (if it exists) become the dominant factors. The conductive network inside the composites stabilizes and becomes hard to change under loading. As a result, the composites have low initial electrical resistivity, and will present more stable sensing properties and low sensing sensitivity.

In general, a low electrical resistivity is desirable in sensing measurements of self-sensing concrete since the low electrical resistivity is helpful for enhancing the signal-to-noise ratio. However, high sensing sensitivity and low filler concentration are difficult to obtain at the same time as low electrical resistivity. Fortunately, there is a balance among the high sensing sensitivity, low filler concentration, and low electrical resistivity near the percolation threshold. Therefore, the percolation threshold is an important parameter for designing and optimizing sensing properties of self-sensing concrete. Generally, the filler concentration above the percolation threshold is beneficial for the sensing sensitivity under tension, while that below the percolation threshold is beneficial for the sensing sensitivity under compression [2,22,28,32].

6.5 Constitutive Model of Sensing Characteristic Behavior

Some constitutive models have been successively developed for verifying the proposed generation mechanism of sensing properties and quantitatively describing sensing properties of self-sensing concrete in recent years. Their modeling principles and application goals are summarized in Table 6.1.

Sett [39] and Garas [40] successfully built a constitutive model by combining the principle of percolation theory and continuum mechanics to predict the self-sensing behavior of polymer concrete and cement mortar with different concentrations of carbon fiber under uniaxial compression (as shown in Figure 6.16), tension, and bending.

TABLE 6.1 Constitutive Sensing Models for Some Typical Self-Sensing Concretes

Type of Self-Sensing Concrete	Loading Mode	Modeling Principle	Goals of Model to Describe or Predict
With carbon fiber [39,40]	Uniaxial compression, tension, and bending	Percolation theory and continuum mechanics	Stress/load sensing behavior of self-sensing concrete with different carbon fiber contents
With carbon fiber [41]	Uniaxial tension and compression	Change in contact electrical resistivity of fiber-matrix interface due to pull-out or pull-in of crack-bridging fiber	Stress sensing behavior
With nickel powder [42]	Uniaxial compression within elastic regime	Field emission effect and interparticle separation change in self-sensing concrete	Stress/strain sensing behavior of self-sensing concrete with different particle sizes of nickel powder
With carbon fiber [18]	Uniaxial compression	Ohmic continuum conduction and tunneling conduction	Strain sensing behavior
With carbon black [43]	Uniaxial compression	Tunneling effect	Strain sensing behavior

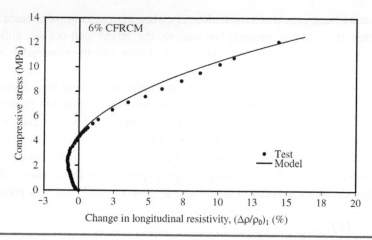

Figure 6.16   Curves of measured and calculated values of the change in electrical resistivity versus compressive stress [40].

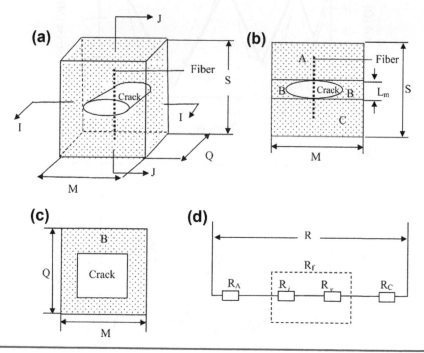

Figure 6.17   (a) Three-dimensional schematic illustration of a parallelepiped unit cell of size $M \times Q \times S$ and containing a single fiber (dotted vertical line) that bridges a crack (white region) in the horizontal plane. The cement matrix is the shade region. (b) J–J section view of the unit cell. (c) I–I section view of the unit cell. (d) Equivalent electrical circuit of the unit cell, showing resistors in series [41].

Wen and Chung provided a model for the carbon fiber cement paste based on the notion that the sensing property is due to the slight pull-out or pull-in of crack-bridging fibers during crack opening or closing and the consequent change in the contact electrical resistivity of the fiber–matrix interface (as shown in Figure 6.17). Good agreement is found between the theory and the experimental results obtained under uniaxial tension and compression (as shown in Figure 6.18).

Han et al. established a constitutive model relating the change in the electrical resistivity of the cement paste containing nickel powder with needle-like surfaces to the applied compressive stress. This model incorporates the field emission effect with the interparticle separation change of nickel powders in composites within the elastic regime under uniaxial compression. It is successfully used to predict the

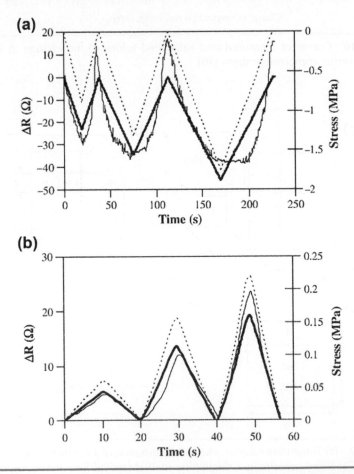

Figure 6.18 Curves of measured (light solid line) and calculated (dark solid line) values of the change in electrical resistance versus time and measured stress (dotted line). (a) Under uniaxial compression. (b) Under uniaxial tension [41].

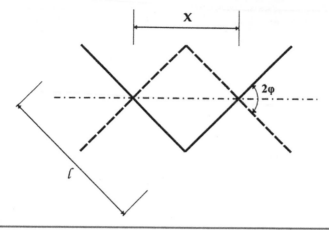

Figure 6.19 Schematic illustration of rotational contacting probability of two carbon fibers in a plane [18].

sensing behavior of the cement paste containing nickel powders with different particle sizes [42]. Xu et al. presented a model for quantitative analysis of the sensing mechanism for the carbon fiber cement mortar under uniaxial compression. The model is based on the concept that the electronic conduction dominates when the composite is in dry state and is a combination of ohmic continuum conduction and tunnel transmission conduction (as shown in Figures 6.19 and 6.20). This model is successfully applied in simulating the strain-sensing characteristics of the composite (as shown in Figure 6.21) [18].

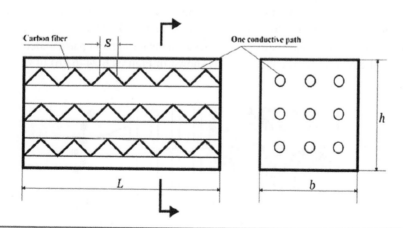

Figure 6.20 Schematic illustration of conductive paths formed by contacting carbon fibers in the longitudinal and cross-sectional directions [18].

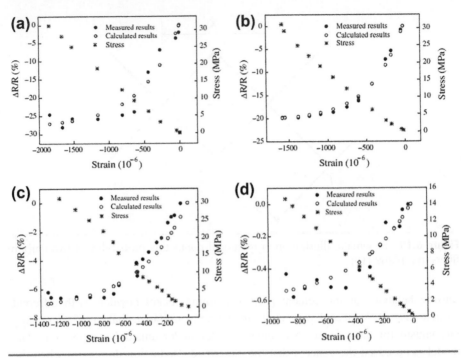

Figure 6.21 Comparison of measured and calculated values of fractional resistance versus strain and stress values versus strain at different fiber volume factions. (a) 0.05%. (b) 0.15%. (c) 0.25%. (d) 0.40% [18].

Xiao et al. proposed a tunnel effect theory-based sensing model to predict strain-sensing property of the carbon black cement paste when carbon black concentration is near the percolation threshold and conductive mechanism is dominated by the tunneling effect (as shown in Figures 6.22 and 6.23). The proposed model is able to predict the resistance behavior of the carbon black cement paste under various loading and environmental conditions (as shown in Figures 6.24–6.26) [43].

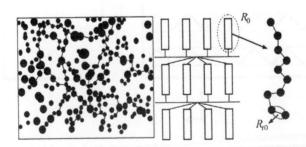

Figure 6.22 Schematic of conductive network in concrete with carbon black [43].

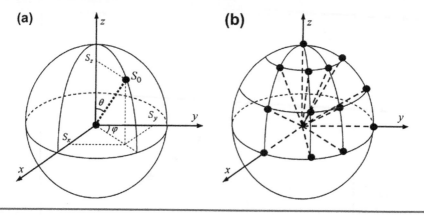

Figure 6.23 Schematic of (a) orientation and (b) distribution of R_{t0} [43].

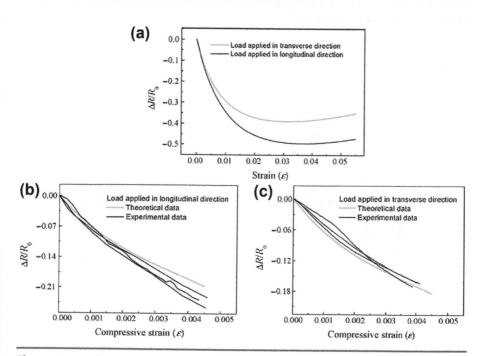

Figure 6.24 Theoretical sensing results of composite under uniaxial compressive loading in various directions (a), and experimental results and theoretical results in limited strain rage (b, c) [43].

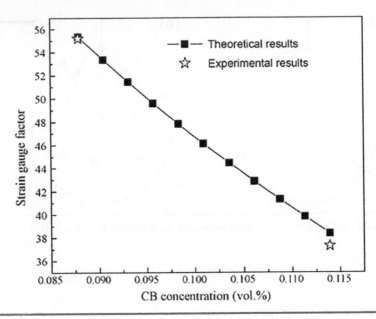

Figure 6.25 Modeling and experimental results of the effects of carbon black concertration on strain gauge factors [43].

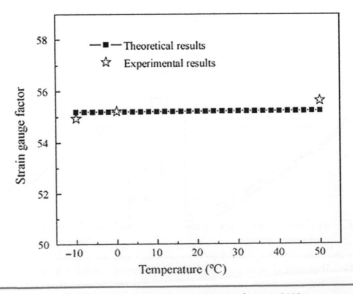

Figure 6.26 Effect of temperature on the strain gauge factors [43].

6.6 Summary and Conclusions

The electrical conduction mechanism of self-sensing concrete is very complex in nature. There are mainly two types of electrical conduction: electronic conduction from conductive functional fillers and ionic conduction from the concrete matrix. Furthermore, the electrical conduction consists of contacting, tunneling conduction, and/or field emission. These conduction types coexist in self-sensing concrete and interrelate with each other. Their contribution on electrical conductivity of the composites closely depends on the concentration of functional fillers. This phenomenon can be described by the percolation theory. Under external force or deformation, the change in electrical resistivity can be attributed to several factors such as change of intrinsic resistance of functional fillers, change of bonding between functional filler and matrix, change of contact between functional fillers, change of tunneling distance between functional fillers, and change of capacitance. These factors work together for contributing sensing property of the composites, but only one or several of them are leading at certain zones of the percolation curve. The percolation threshold is an important parameter to design and optimize the sensing properties of the composites. The mechanism provides the principle for establishment of the constitutive models. The effective constitutive models may help to not only verify the mechanism of sensing properties but also describe and predict sensing properties of self-sensing concrete.

References

[1] Mao QZ, Zhao BY, Sheng DR, Li ZQ. Resistance changement of compression sensible cement specimen under different stresses. J Wuhan Univ Technol 1996;11:41–5.
[2] Li H, Xiao HG, Ou JP. Effect of compressive strain on resistance of carbon black filled cement-based composites. Cem Concr Compos 2006;28:824–8.
[3] Li GY, Wang PM, Zhao XH. Pressure-sensitive properties and microstructure of carbon nanotube reinforced cement composites. Cem Concr Compos 2007;29:377–82.
[4] Sun MQ, Li ZQ, Liu QP. The electromechanical effect of carbon fiber reinforced cement. Carbon 2000;40(12):2263–4.
[5] Han BG. Properties, sensors and structures of pressure-sensitive carbon fiber cement paste [Dissertation for the Doctor Degree in Engineering]. China: Harbin Institute of Technology; 2005.
[6] Wang YL, Zhao XH. Positive and negative pressure sensitivities of carbon fiber-reinforced cement-matrix composites and their mechanism. Acta Mater Compos Sin 2005;22(4):40–6.
[7] Han BG, Yu X, Ou JP. Chapter 1: multifunctional and smart carbon nanotube reinforced cement-based materials. In: Gopalakrishnan K, Birgisson B, Taylor P, Attoh-Okine NO, editors. Book: nanotechnology in civil infrastructure: a paradigm shift. Publisher: Springer; 2011. pp. 1–47 (276).
[8] Li CT, Qian JS, Tang ZQ. Study on properties of smart concrete with steel slag. China Concr Cem Prod 2005;2:5–8.

[9] Han BG, Ou JP. Embedded piezoresistive cement-based stress/strain sensor. Sensors Actuators Physical 2007;138(2):294–8.

[10] Zhang ZM. Nano/microscale heat transfer. New York: McGraw-Hill; 2007.

[11] Wan Y, Wen DJ. Conducting polymer composites and their special effects. Chin J Nat 1999;21(3):149–53.

[12] Chen K, Xiong CX, Li LB, Zhou L, Lei Y, Dong LJ. Conductive mechanism of antistatic poly (ethylene terephthalate)/ZnOw composites. Polym Compos 2008;30:226–31.

[13] Celzard A, Mcrae E, Furdin G, Mareche JF. Conduction mechanisms in some graphite- polymer composites: the effect of a direct-current electric field. J Phys Condens Matter 1997;9:2225–37.

[14] Peng N, Zhang Q, Lee YC, Huang H, Tan OK, Tian JZ, et al. Humidity and temperature effects on carbon nanotube field-effect transistor-based gas sensors. Sens Lett 2008;6(6):796–9.

[15] Han BG, Yu X, Ou JP. Effect of water content on the piezoresistivity of CNTs/cement composites. J Mater Sci 2010;45:3714–9.

[16] Han BG, Han BZ, Yu X, Ou JP. Ultrahigh pressure-sensitive effect induced by field emission at sharp nano-tips on the surface of spiky spherical nickel powders. Sens Lett 2011;9(5):1629–35.

[17] Wittmann FH. Materials for buildings and structures. Wiley-VCH; 2000.

[18] Xu J, Zhong WH, Yao W. Modeling of conductivity in carbon fiber-reinforced cement-based composite. J Material Sci 2010;45:3538–46.

[19] Wen SH, Chung DDL. The role of electronic and ionic conduction in the electrical conductivity of carbon fiber reinforced cement. Carbon 2006;44:2130–8.

[20] Wen SH, Chung DDL. Electric polarization in carbon fiber-reinforced cement. Cem Concr Res 2001;31:141–7.

[21] Zheng LX, Song XH, Li ZQ. Investigation on the method of AC measurement of compression sensivility of carbon fiber cement. J Huazhong Univ Sci Technol Urban Sci Ed 2005;22(2):27–9.

[22] Wang XF, Wang YL, Jin ZH. Electrical conductivity characterization and variation of carbon fiber reinforced cement composite. J Mater Sci 2002;37:223–7.

[23] Xie P, Gu P, Beaudoin JJ. Electrical percolation phenomena in cement composites containing conductive fibres. J Mater Sci 1996;31:4093–7.

[24] Huang Y, Xiang B, Ming XH, Fu XL, Ge YJ. Conductive mechanism research based on pressure-sensitive conductive composite material for flexible tactile sensing. In: Proceeding of the 2008 IEEE international conference on information and automation (Zhangjiajie); 2008. pp. 1614–9.

[25] Mohammed HA, Uttandaraman S. A review of vapor grown carbon nanofiber/polymer conductive composites. Carbon 2009;47(1):2–22.

[26] Chen B, Wu KR, Yao W. Conductivity of carbon fiber reinforced cement-based composites. Cem Concr Compos 2004;26:291–7.

[27] Wu SP, Mo LT, Shui ZH, Chen Z. Investigation of the conductivity of asphalt concrete containing conductive fillers. Carbon 2005;43:1358–63.

[28] Saafi M. Wireless and embedded carbon nanotube networks for damage detection in concrete structures. Nanotechnology 2009;20:395502 (7pp).

[29] Hwang SH, Park YB, Yoon KH, Bang DS. Chapter 18 Smart materials and structures based on carbon nanotube composites. In: Yellampalli Siva, editor. Book: carbon nanotubes-synthesis, characterization, applications. Publisher: InTech; 2011. pp. 371–96 (514p).

[30] Jiang CX, Li ZQ, Song XH, Lu Y. Mechanism of functional responses to loading of carbon fiber reinforced cement-based composites. J Wuhan Univ Technology-Mater Sci Ed 2008;23(4):571–3.

[31] Tombler TW, Zhou C, Alexseyev L, Kong J, Dai H, Liu L, et al. Reversible electromechanical characteristics of carbon nanotubes under local-probe amanipulation. Nature 2000;405:769–72.

[32] Pushparaj VL, Nalamasu O, Manoocher Birang M. Carbon nanotube-based load cells. Patent US2010/0050779 A1, 2010.

[33] Han BG, Han BZ, Yu X. Effects of content level and particle size of nickel powder on the piezoresistivity of cement-based composites/sensors. Smart Mater Struct 2010;19:065012 (6pp).

[34] Chung DDL. Piezoresistive cement-based materials for strain sensing. J Intelligent Material Syst Struct 2002;13(9):599–609.

[35] Fu XL, Chung DDL. Contact electrical resistivity between cement and carbon fiber: its decrese with increasing bond strenght and its increase during fiber pull-out. Cem Concr Res 1995;25(7):1391–6.

[36] Fu XL, Chung DDL. Sensitivity of the bond strength to the structure of the interface between reinforcement and cement, and the variability of this structure. Cem Concr Res 1998;28(6):787–93.

[37] Hou TC, Lynch JP. Conductivity-based strain monitoring and damage characterization of fiber reinforced cementitious structural components. Proceeding SPIE 2005;5765:419–29.

[38] Han BG, Zhang K, Yu X, Kwon E, Ou JP. Electrical characteristics and piezoresistive response measurements of carboxyl MWNT/cement composites. Cem Concr Compos 2012;34:794–800.

[39] Sett K. Characterization and modeling of structural and self-monitoring behavior of fiber reinforced polymer concrete [Dissertation for the Master of Science in Civil Engineering]. USA: University of Houston; 2003.

[40] Garas VY. Characterization and modeling of structural and self-monitoring behavior of carbon fiber reinforced cement mortar [Dissertation for the Master Degree in Science in Civil Engineering]. Huston (USA): University of Huston; 2004.

[41] Wen SH, Chung DDL. Model of piezoresistivity in carbon fiber cement. Cem Concr Res 2006;36(10):1879–85.

[42] Han BG, Han BZ, Yu X, Ou JP. Piezoresistive characteristic model of nickel/cement composites based on field emission effect and inter-particle separation. Sens Lett 2009;7(6):1044–50.

[43] Xiao HG, Li H, Ou JP. Modeling of piezoresistivity of carbon black filled cement-based composites under multi-axial strain. Sensors Actuators: Phys 2010;160(1–2):87–93.

Chapter 7

Applications of Self-Sensing Concrete

Chapter Outline

7.1 Introduction and Synopsis

The goal of developing self-sensing concrete is to apply it in civil structures. Owing to the capability of detecting its inside stress, strain, cracking, and damage, self-sensing concrete can replace embedded or attached sensors or detectors, which suffer from high cost, low durability, limited sensing volume, and degradation of the **189**

Self-Sensing Concrete in Smart Structures. http://dx.doi.org/10.1016/B978-0-12-800517-0.00007-1

structural performance of the concrete (for embedded sensors or detectors). Therefore, self-sensing concrete has potential for structural health monitoring, traffic detection, and border/military security applications. Currently, application of self-sensing concrete in the fields of structural health monitoring and traffic detection has gained much attention. This chapter will introduce these two major applications of self-sensing concrete.

7.2 Structural Health Monitoring

A civil structure should meet the requirements of safety, durability, serviceability, and sustainability for long-term operation. During its long-term service, the structure may slowly deteriorate in its performance or be severely damaged or even collapse when subjected to natural disasters such as earthquakes and strong winds. Structural health monitoring technology provides a way to evaluate the safety and durability of a structure during its service life, to ensure its serviceability and sustainability. Sensor technology is a critical part of the structural health monitoring system [1]. For concrete structures, resistance strain gauges, optic sensors, piezoelectric ceramic, shape memory alloy, and fiber-reinforced polymer bar have been used as sensors in health monitoring systems. However, these sensors have two main drawbacks: poor durability and unfavorable compatibility with concrete structures [1]. Self-sensing concrete is considered a promising candidate for solving the durability and incompatibility issue between conventional sensors and concrete structures. This is possible because self-sensing concrete itself functions as both the structure and the sensor. By measuring the electrical resistance of the concrete, its structural health and stress, strain, cracking, and damage can be monitored (as shown in Figure 7.1 [2]).

As shown in Figure 7.2, the self-sensing concrete structure for structural health monitoring can be used in bulk, coating, sandwich, bonded, and embedded forms. Compared with the bulk form, the latter four forms can achieve higher monitoring efficiency and lower construction cost, in which self-sensing concrete lies only in key positions of the structures. The advantage of the bulk form against others, especially the coating and sandwich forms, is simple construction technology.

7.2.1 In Bulk Form

Bulk form means that the structure or component is wholly made of self-sensing concrete. Previous research on the application of self-sensing concrete for structural health monitoring in bulk form is summarized in Table 7.1.

Sett first fabricated beams and rings using the polymer concrete with carbon fiber [3]. He tested the relationship between change in electrical resistance and loading/midspan deflection of the beams under four-point bending (as shown in

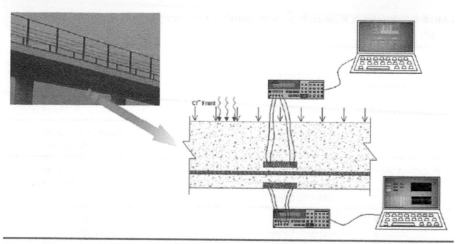

Figure 7.1 Structural health monitoring system of bridge based on self-sensing concrete [2].

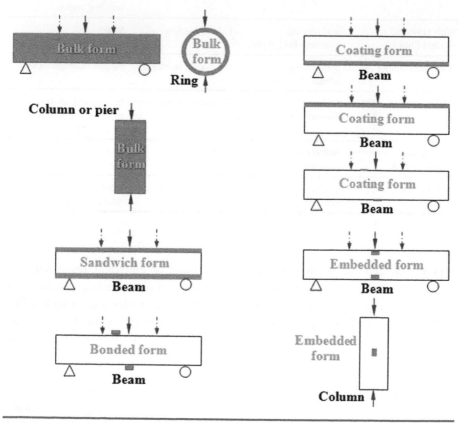

Figure 7.2 Typical application forms of self-sensing concrete for structural health monitoring (parts in red represent self-sensing concrete).

TABLE 7.1 Previous Research on Self-Sensing Concrete in Bulk Form for Structural Health Monitoring

Type of Self-Sensing Concrete	Component	Loading Mode	Parameters to Monitor
With carbon fiber (polymer concrete) [3]	Self-sensing concrete beam	Four-point bending	Load and deflection
	Self-sensing concrete ring	Parallel plate loading	Load and change in ring diameter
With carbon fiber [4]	Reinforced self-sensing concrete beam	Four-point bending	Load and damage in pure bending region
With carbon fiber [5]	Self-sensing concrete beam	Four-point bending	Load and damage in compressive and tensile zones
With carbon fiber [6]	Self-sensing concrete beam	Four-point bending	Elastic compressive stress and strain in the pure bending region
With polyvinyl alcohol fiber [7]	Self-sensing concrete bridge pier	Lateral loading	Strain
With carbon fiber [8]	Self-sensing concrete beam	Four-point bending	Fatigue damage extent
With carbon fiber [9]	Self-sensing concrete beam	Three-point bending	Initial load, elastic deformation, deflection, and fracture
With carbon nanofiber [10]	Self-sensing concrete column	Compression	Strain
With carbon nanotubes [11]	Self-sensing concrete beam	Four-point bending	Central deflection and crack initiation

Figure 7.3) and the relationship between change in electrical resistance and loading/change in diameter of the rings under parallel plate loading (as shown in Figure 7.4), respectively. Test results for both the beams under four-point bending and the rings under parallel plate loading showed that the applied load or the deflection resulting from that load can be predicted by monitoring the change in resistance (as shown in Figure 7.5) [12].

Zhang et al. fabricated reinforced concrete beams with intrinsic self-sensing carbon fiber concrete. A four-point bending test indicates that the beams have the capability of sensing load and damage in their pure bending region [4]. Zhang et al. investigated variations in the electrical resistance of carbon fiber concrete

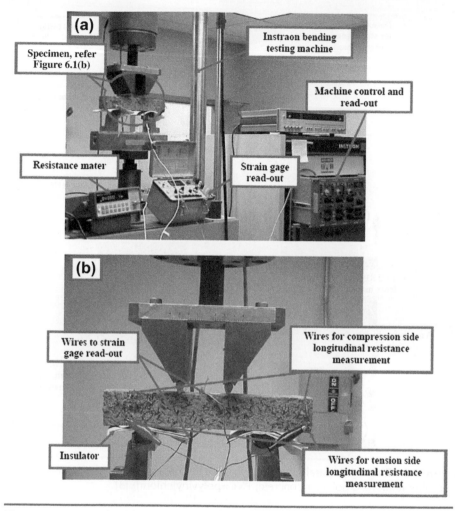

Figure 7.3   Four-point bending test (a) test up and (b) specimen [3].

beams under monotonous, repeated, and alternating bending loadings. They observed that carbon fiber concrete beams can monitor load and damage in compressive and tensile zones under four-point bending [5]. Zhang ct al. studied the feasibility of elastic compressive stress self-monitoring in the pure bending region of carbon fiber concrete beams under four-point bending. The results of a cyclic loading experiment showed that elastic stress self-monitoring of the key section of reinforced concrete beams can be achieved based on the correlation between electrical resistance and the elastic compression strain (as shown in Figure 7.6) [6].

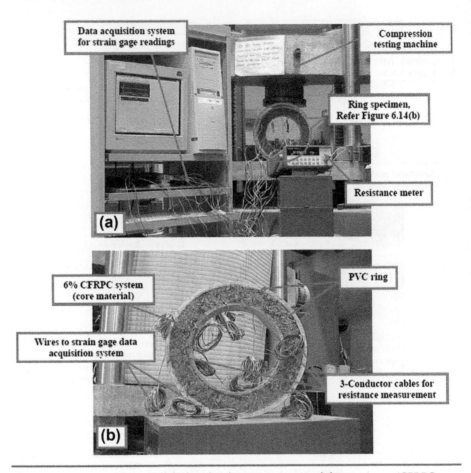

Figure 7.4 Ring under parallel plate loading (a) test up and (b) specimen (CFRPC=carbon fiber reinforced polymer concrete; PVC=polyvinylchlorid) [3].

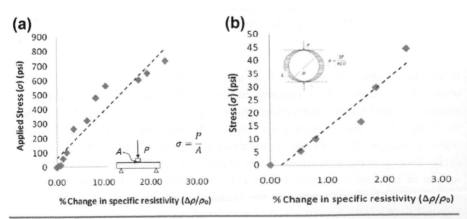

Figure 7.5 Variation of stress with resistivity during (a) bending loading and (b) tensile loading [12].

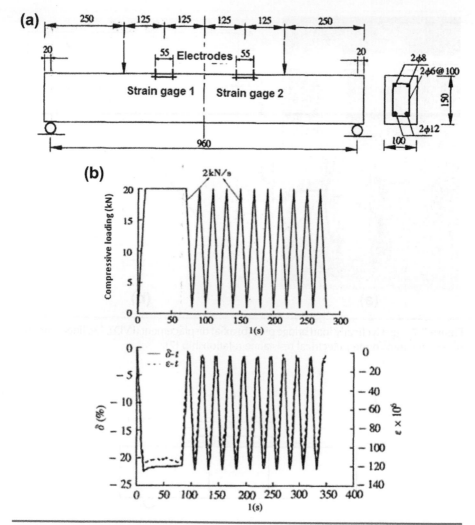

Figure 7.6 Fractional change in electrical resistance (δ) of concrete with carbon fiber versus elastic compression strain (ε) of beam [6]. (a) Diagram of beam structure, (b) piezoresistive response.

Hou and Lynch used fiber-reinforced engineered cementitious composites to fabricate a one-third scaled bridge pier (as shown in Figure 7.7(a)) and tested the self-sensing behavior of the bridge pier laterally loaded under cyclically repeated drift reversals [7]. The test results revealed the feasibility of using fiber-reinforced engineered cementitious composites to self-sense the strain of the bridge pier (as shown in Figure 7.7(b)) [7].

Wang et al. tested four-point bending beams fabricated with carbon fiber concrete and evaluated the relationship between electrical property and fatigue life under

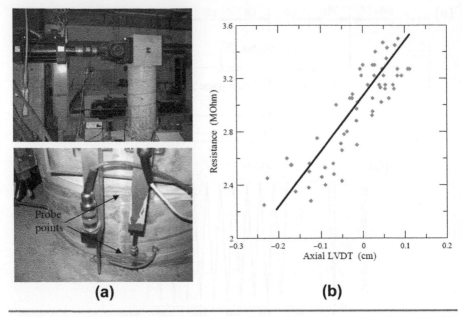

(a) **(b)**

Figure 7.7   (a) Cyclically load bridge pier, (b) axial displacement (LVDT, i.e. linear variable differential transformer)-electrical resistance relationship [7].

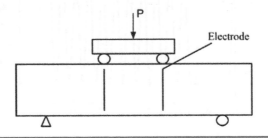

Figure 7.8   Diagram of beam [8].

cyclic flexural loading. The test results indicated that carbon fiber concrete can be used to monitor the extent of fatigue damage and predict their fatigue life (as shown in Figures 7.8–7.10) [8].

Zhang et al. fabricated carbon fiber concrete beams and tested the self-sensing behavior of the beams under three-point bending [9]. They observed that carbon fiber concrete can be used to attain the initial load. The self-sensing property of carbon fiber concrete can be used to explain Mode I fracture of three-point bending concrete beams. The relationships between electrical resistance and load, deflection, and crack mouth opened deflection are in accord with Lorentz fitting. The relative

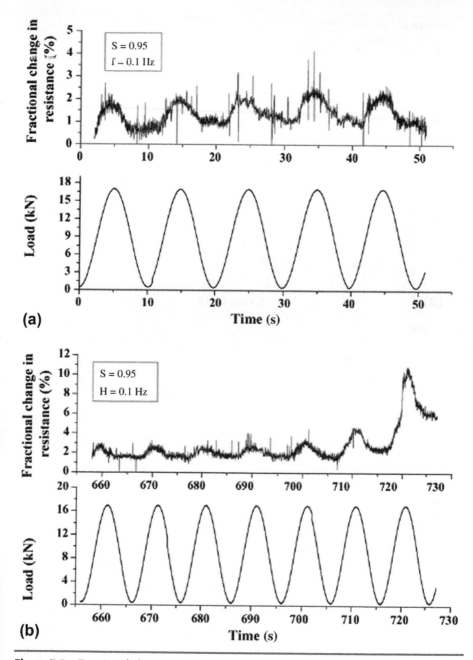

Figure 7.9   Fractional change in electrical resistance during repeated flexural loading at S (stress ratio) = 0.95: (a) first five cycles and (b) last six cycles.

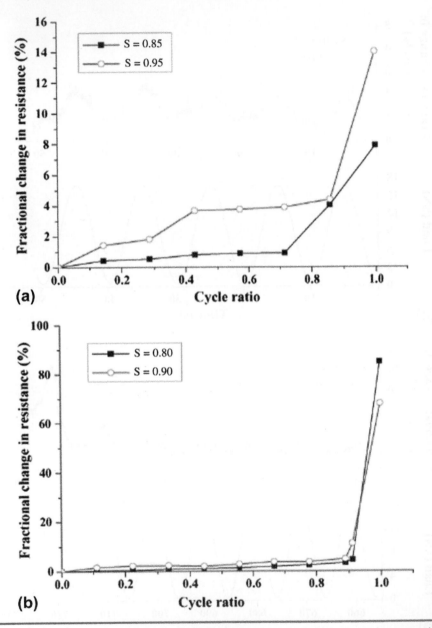

Figure 7.10 Fractional change in electric resistance vs cycle ratio (ratio between the individual testing cycle number and the total number of cycles to failure under cyclic flexural loading) under different stress ratios (S): (a) before fracture and (b) until fracture [8].

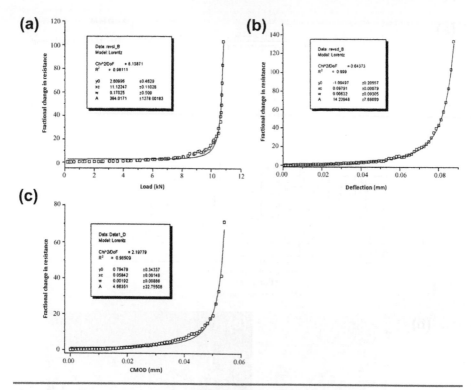

Figure 7.11 Structural self-sensing behavior of beams fabricated with self-sensing concrete with carbon fiber [9]. (a) Relation between relative variation of resistance and load, (b) Relation between relative variation of resistance and deflection, (c) Relation between relative variation of resistance and crack mouth opened deflection (CMOD).

change curves of resistance can describe the three stages of elastic deformation, stable fracture, and unstable fracture of the beams before maximum load (as shown in Figure 7.11) [9].

Howser et al. built shear-critical columns with the carbon nanofiber concrete and tested the self-sensing behavior of concrete columns under a reversed cyclic load. The columns were capable of monitoring their own strain (as shown in Figure 7.12) [10].

Saafi et al. fabricated beams with self-sensing concrete with carbon nanotubes, which were subjected to four-point bending tests for sensing characterization in the tension side of the beams (as shown in Figure 7.13(a)) [11]. They observed that the change in the complex impedance was linearly proportional to the change in the central deflection with a rapid nonlinear increase near the failure load. The electrical impedance of the beams tended to gradually increase with the applied load. Crack propagation at the mid-span of the beams caused a sudden increase in the measured electrical impedance. This sensing mechanism can be used as a tool

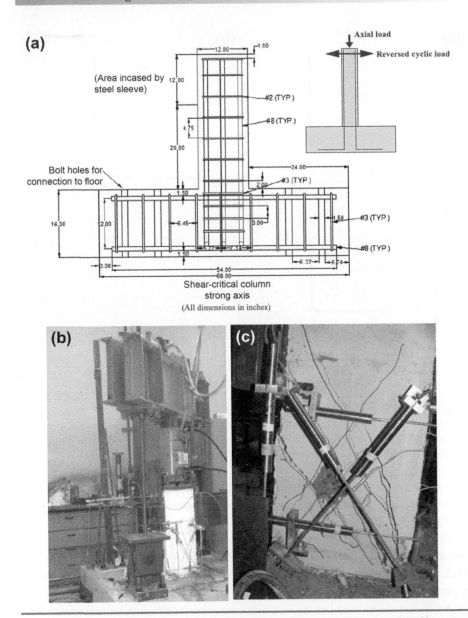

Figure 7.12 Structural self-sensing behavior of columns fabricated with self-sensing concrete with carbon nanofiber [10]. (a) Elevation view of the strong axis of the shear-critical columns and foundations, (b) Axial loading system, (c) West side of column after failure, (d) Electrical resistance variation (ERV) versus horizontal deflection at top.

(d)

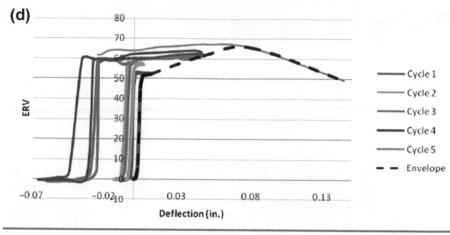

Figure 7.12   *(continued)*

to detect crack initiation and propagation in concrete structures (as shown in Figure 7.13(b)) [11].

7.2.2 IN COATING FORM

Coating form means that one surface of a structure or component is covered with a layer of self-sensing concrete. Previous research on the application of self-sensing concrete for structural health monitoring in coating form is summarized in Table 7.2.

Wen and Chung first fixed carbon fiber cement paste coating on either the tensile side or the compressive side of cement paste beams under three-point bending (as shown in Figure 7.14(a)) [13]. They observed that the carbon fiber cement paste coating can be used for compressive or tensile strain monitoring of the beam surface (as shown in Figure 7.14(b) and (c)) [13].

Huang et al. configured carbon fiber concrete layers at the bottom of concrete beams and reinforced concrete beams, and tested real-time monitoring performance of the beams under four-point bending [14]. They observed that the concrete beams could monitor real-time loading and strain. The reinforced concrete beams could give pre-warning of fracture, ultimate load, and rigidity (as shown in Figure 7.15 and Table 7.3) [14].

Zhang et al. laid carbon fiber concrete layers at the top or bottom of concrete beams and tested the self-sensing behavior of the beams under monotonous and repeated four-point bending [15]. Test results indicate that the beams with carbon fiber concrete layers were capable of monitoring their loading, deformation, and degree of damage. Hong made reinforced concrete beams with graphite slurry

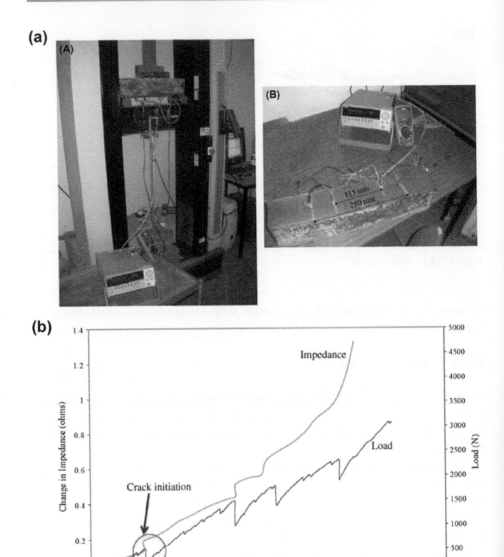

Figure 7.13 Structural self-sensing behavior of beams fabricated with self-sensing concrete with carbon nanotubes [11]. (a) Experimental setup: (A) loading test; (B) test for electrical property, (b) change in complex impedance vs deflection response of geopolymeric concrete beams with 0.5 wt.% MWCNTs.

TABLE 7.2 Previous Research on Self-Sensing Concrete in Coating Form for Structural Health Monitoring

Type of Self-Sensing Concrete	Component	Loading Mode	Parameters to Monitor
With carbon fiber [13]	Concrete beam cast self-sensing concrete on its top or bottom	Three-point bending	Compressive or tensile strain of beam surface
With carbon fiber [14]	Concrete/reinforced concrete beam cast self-sensing concrete on its bottom	Four-point bending	Load and strain
With carbon fiber [15]	Concrete beam cast self-sensing concrete on its top or bottom	Four-point bending	Load, deformation, and degree of damage
With steel fiber [16]	Reinforced concrete beam cast self-sensing concrete on its top or bottom	Four-point bending	Damage
With carbon fiber [17]	Concrete beam cast self-sensing concrete on its bottom	Four-point bending	Damage condition and extent of fatigue damage
With carbon fiber or carbon nanofiber [18]	Reinforced concrete beam cast self-sensing concrete on its top, bottom or side	Four-point bending	Strain

infiltrated steel fiber concrete coating on their top or bottom, and studied the variation in electrical resistance of the beams under cyclic four-point bending at different stress amplitudes [16]. The internal damage of the beams could be monitored by measuring the variation in electrical resistance of the graphite slurry infiltrated steel fiber concrete coating. Wang et al. performed an experiment testing four-point bending beams fabricated with the common reinforced concrete beam with a layer of carbon fiber concrete on their bottom (as shown in Figure 7.16(a) and (b)) [17] and analyzed the electrical property of the designed beams under monotonic loading as well as the relationship between electrical property and fatigue damage under cyclic flexural loading. Damage and the extent of fatigue damage of the beams could be monitored using carbon fiber concrete as a sensing coating (as shown in Figure 7.16(c)–(g)) [17].

Baeza et al. cast self-sensing concrete with carbon fiber or carbon nanofiber on the top or bottom in the pure bending region of reinforced concrete beams subjected to four-point bending (as shown in Figures 7.17 and 7.18) [18]. They observed that self-sensing concrete with carbon fiber or carbon nanofiber was capable of acting as a strain sensor attached to a reinforced concrete beam, even if the structure was close

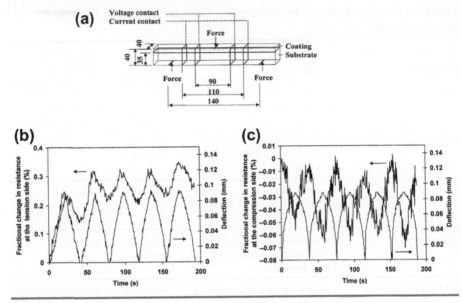

Figure 7.14 Structural self-sensing behavior of beam cast self-sensing concrete containing carbon fiber on its top or bottom [13]. (a) Specimen configuration. The configuration shown is for the case of the coating at the compression side of the flexural specimen, (b) fractional change in resistance and deflection during cyclic flexure for the case in which the strain-sensing coating is carbon fiber concrete at the tension side, (c) fractional change in resistance and deflection during cyclic flexure for the case in which the strain-sensing coating is carbon fiber concrete at the compression side.

to collapse, but it was not applicable as a damage sensor owing to the low strains that concrete experiences (as shown in Figure 7.19) [18].

7.2.3 IN SANDWICH FORM

Sandwich form means that the top and bottom surfaces of a structure or component are both covered with self-sensing concrete layers. Previous research on the application of self-sensing concrete for structural health monitoring in sandwich form is summarized in Table 7.4.

Zheng et al. laid carbon fiber concrete layers on both the top and bottom of concrete beams and tested the self-sensing performance of sandwich concrete beams under three-point bending (as shown in Figure 7.20) [19]. Test results within the elastic stage indicated that the sandwich concrete beams were capable of stress and strain monitoring of the compressive and tensile zones.

Hong used graphite slurry infiltrated steel fiber concrete as sensing layers to fabricate sandwich concrete beams and tested the self-sensing behavior of the beams under four-point bending (as shown in Figure 7.21(a)) [20]. The corresponding

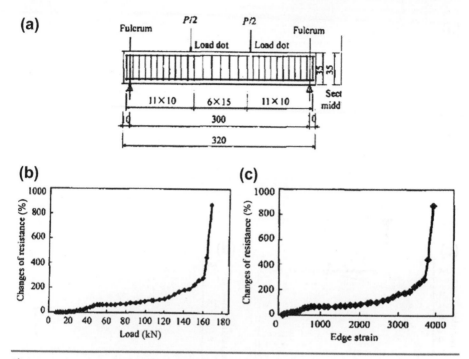

Figure 7.15 Structural self-sensing behavior of beam cast self-sensing concrete containing carbon fiber on its bottom [14]. (a) Schematic diagram of beam, (b) electrical resistivity versus load, (c) electrical resistivity versus strain.

TABLE 7.3 Electrical Resistivity Variation Comparison between Theoretical and Experimental Values [14]

Pre-warn Clicking	Fracture	Limit Load	Rigidity
Theoretical value	6.51	534.80	65.30
Experimental value	7.53	565.32	69.51

relationships between electrical resistance, loading, deflection, and strain indicated that the graphite slurry infiltrated steel fiber concrete layers were capable of real-time health monitoring of compressive and tensile regions of the sandwich concrete beams (as shown in Figure 7.21(b)–(d)) [20].

Wu et al. cast carbon fiber concrete layers onto the top and bottom of reinforced concrete beams and investigated the variations in electrical resistance of the carbon fiber concrete layers under four-point bending [21]. Relationships between electrical resistance, loading, deflection, and cracks showed that the change in electrical resistance could be used to monitor the extent of damage in the designed beams (as shown in Figure 7.22(a)–(d)) [21].

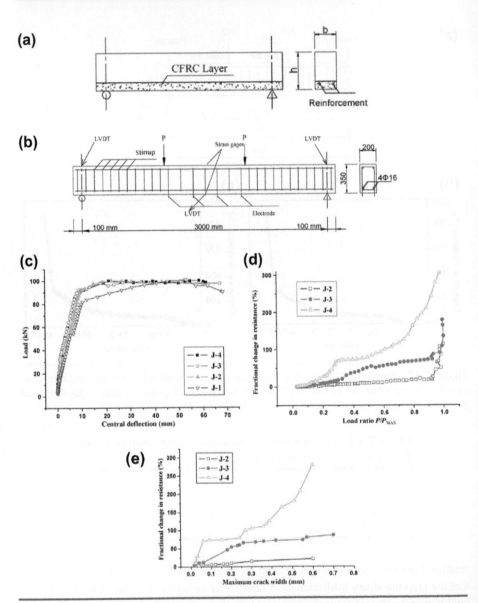

Figure 7.16 Structural self-sensing behavior of reinforced concrete beam cast self-sensing concrete containing carbon fiber on its bottom [17]. (a) Schematic diagram of reinforced concrete beam, (b) details of the test beam, (c) load vs deflection curve for different groups of beams, (d) fractional change in electrical resistance vs load ratio, (e) fractional change in electrical resistance vs maximum cracking, (f) fractional change in resistance during repeated flexural loading at first five cycles, (g) Fractional change in resistance during repeated flexural loading at last five cycles.

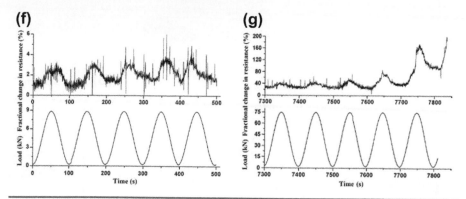

Figure 7.16 *(continued)*

Chen and Liu cast the carbon fiber cement mortar on the upper (compressive) face and the lower (tensile) face of cement mortar specimens [22]. The correlations between relative resistance change and deflection curves under three-point bending are shown in Figure 7.23. When the carbon fiber-cement mortar was cast on the face under compression, the resistance increased with the increase in deflection during loading, which was also reversible in each cycle, whereas when the carbon fiber-cement mortar was cast on the face under tension, resistance decreased with the increase in deflection. The correlations between relative resistance change and deflection curves under three-point bending indicated that the coating could effectively monitor the loading process.

7.2.4 IN EMBEDDED FORM

Embedded form means that self-sensing concrete is prefabricated into standard small-size sensors and then is embedded into the structure. Previous research on the application of self-sensing concrete for structural health monitoring in embedding form are summarized in Table 7.5.

Ou and Han first embedded cement paste sensors with hybrid carbon fiber and carbon black into compressive zone of the pure bending region of reinforced concrete beams and into the center of concrete columns, and tested the self-sensing performance of these components under four-point bending and uniaxial compression [23]. They found that these components embedded with cement paste sensors were capable of compressive strain self-sensing.

Chacko studied electrical resistivity versus the load behavior of carbon fiber cement paste sensors embedded in concrete cylinders and beams under compressive and flexural loading [24]. The sensors might more accurately monitor strain in the concrete members in their electrical resistivity reading if they were made of resilient material with high yield strength and low modulus of elasticity. In addition, sensors

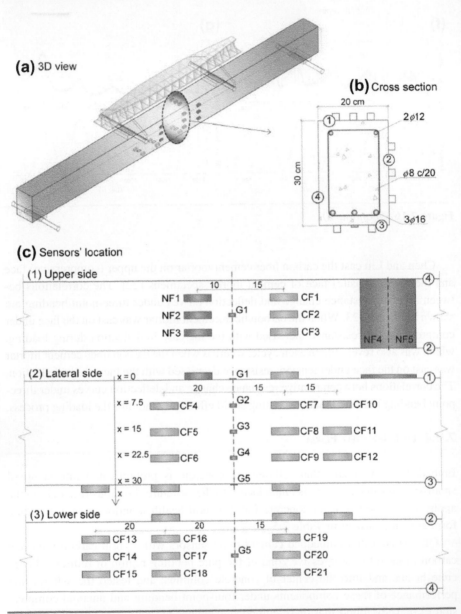

Figure 7.17 (a) Three dimensional (3D) view of the distribution of self-sensing concrete; also, loading and support conditions are represented; (b) cross-section dimensions and steel rebar's arrangement; (c) location and nomenclature of self-sensing concrete (NF = carbon nanofiber; CF = carbon fiber; G = strain gage) [18].

Figure 7.18   Application form of self-sensing concrete: (a) bonded, (b) coating [18].

placed on the downside of beams may perform well as crack sensors. Azhari embedded sensors fabricated with cement paste containing 15% carbon fiber into the bottom side of flexural beams (as shown in Figure 7.24) [2]. The change in electrical resistance of the sensors during cyclic flexural loading of the beams indicated that the sensors were subjected to compression rather than tension. This probably resulted from insufficient bonding. The response of the sensors was relatively compatible with the loading cycles. The author suggested that with a proper embedding procedure and adequate bonding, the sensors could be effectively used to sense stress/strain and cracking of a structure under flexural loading (as shown in Figure 7.25) [2].

Xiao et al. centrally embedded concrete with carbon black into concrete columns with different compressive strengths (as shown in Figure 7.26) and tested the self-monitoring ability of the columns under cyclic load and monotonic load [25]. The concrete columns embedded with carbon black–based self-sensing concrete detected the strain (as shown in Figures 7.27 and 7.28) [25]. Xiao et al. also embedded concrete with carbon black into three different stress zones (i.e., uniaxial compression, combined compression and shear, and uniaxial tension zones) of bending reinforced concrete beams (as shown in Figure 7.29) and investigated the strain sensing properties of the concrete beams under four-point bending [26]. All sensors embedded under the three different stress zones exhibited reasonable sensing results (as shown in Figures 7.30–7.32) [26].

Fan et al. embedded concrete with carbon fiber and graphite into concrete columns that were subjected to different amplitudes of cyclic loading [27]. The electrical resistance was correlated with the force applied on the concrete columns. Saafi embedded single-walled carbon nanotube cement paste into the tensile region of concrete beams to set up self-sensing concrete structures subjected to monotonic and cyclic bending (as shown in Figures 7.33 and 7.34) [28]. The self-sensing concrete

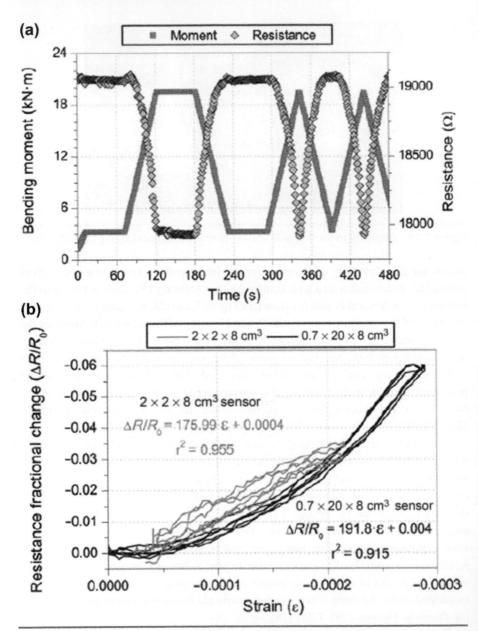

Figure 7.19 Test results: (a) electrical resistance and bending moment in the middle cross-sections vs time; (b) resistance fractional change vs longitudinal strain curves (black curve) [18].

TABLE 7.4 Previous Research on Self-Sensing Concrete in Sandwich Form for Structural Health Monitoring

Type of Self-Sensing Concrete	Component	Loading Mode	Parameters to Monitor
With carbon fiber [19]	Concrete beam cast self-sensing concrete on its top and bottom	Three-point bending	Stress/strain of compressive and tensile zones within elastic stage
With steel fiber [20]	Concrete beam cast self-sensing concrete on its top and bottom	Four-point bending	Loading, deflection and strain of compressive and tensile region
With carbon fiber [21]	Reinforced concrete beam cast self-sensing concrete on its top and bottom	Four-point bending	Load, deflection, crack and damage extent
With carbon fiber [22]	Concrete beam cast self-sensing concrete on its top and bottom	Three-point bending	Loading process and deflection

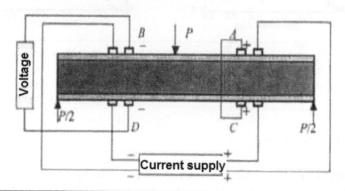

Figure 7.20   Concrete beam cast self-sensing concrete with carbon fiber on its top and bottom [19].

structures could detect crack propagation and damage accumulation during loading by using the change in electrical resistance of the embedded single-walled carbon nanotube cement paste (as shown in Figures 7.35) [28].

Baeza et al. embedded self-sensing concrete with carbon fiber or carbon nanofiber into different service locations of reinforced concrete beams (as shown in Figure 7.17) [18]. They observed that self-sensing concrete with carbon fiber or carbon nanofiber could achieve elastic strain monitoring of compression and tension regions of the beam (as shown in Figure 7.36) [18].

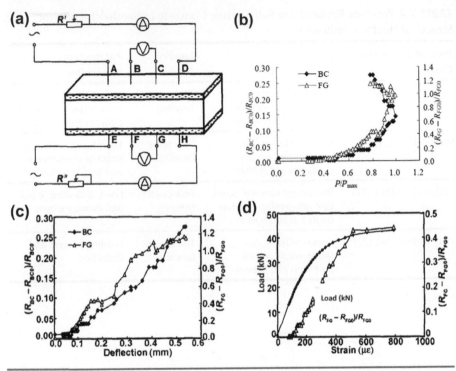

Figure 7.21 Structural self-sensing behavior of beam cast self-sensing concrete with steel fiber on its top and bottom [20]. (a) Schematic diagram of beam, (b) curve of relative variations of resistance with loading, (c) fractional resistance vs deflection, (d) relationship between load, fractional resistance, and strain in the tensile part.

7.2.5 IN BONDED FORM

Bonded form means that small sensors made of self-sensing concrete were attached to the concrete component using glue bonding materials and such. Baeza et al. bonded self-sensing concrete containing carbon fiber or carbon nanofiber on the top, bottom, and side of reinforced concrete beam (as shown in Figures 7.17 and 7.18) [18]. They observed that self-sensing concrete with carbon fiber or carbon nanofiber is capable of measuring strains on the surface of a structural element, regardless of whether the local stresses were tension or compression. Furthermore, almost no difference was found between gauge factors calculated for self-sensing concrete with carbon fiber (with equal dimensions) located on the upper and lower sides of the same beam's section. The gauge factor of self-sensing concrete with carbon nanofiber is around 190. Hence, these composites could act as strain sensors, even for severely damaged structures near collapse (as shown in Figures 7.37–7.39) [18].

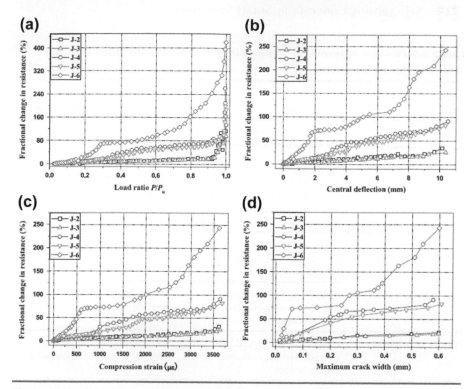

Figure 7.22 Structural self-sensing behavior of reinforced concrete beam cast self-sensing concrete with carbon fiber on its top and bottom [21]. (a) Fractional change in electrical resistance versus load ratio for different groups of test beams, (b) fractional change in electrical resistance versus deflection for different groups of test beams, (c) fractional change in electrical resistance versus compression strain for different groups of test beams, (d) fractional change in electrical resistance versus crack width for different groups of test beams.

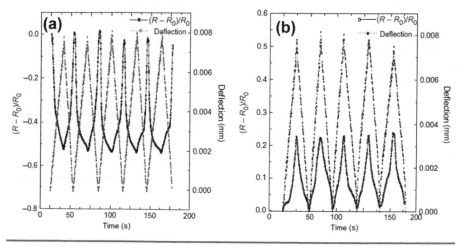

Figure 7.23 Fractional change in resistance and deflection during cyclic flexure: (a) under compression, (b) under tension [22].

TABLE 7.5 Previous Research on Self-Sensing Concrete in Embedded Form for Structural Health Monitoring

Type of Self-Sensing Concrete	Component	Loading Mode	Parameters to Monitor
With hybrid carbon fiber and carbon black [23]	Reinforced concrete beam embedded with self-sensing concrete in its compressive zone of the pure bending region	Four-point bending	Stress and strain
	Concrete column embedded with self-sensing concrete into its center	Uniaxial compression	
With carbon fiber [24]	Concrete cylinder embedded with self-sensing concrete along its longitudinal axis	Compression	Strain and crack
	Concrete beam embedded with self-sensing concrete into its bottom	Four-point bending	
With carbon fiber [2]	Concrete beam embedded with self-sensing concrete into its bottom	Four-point bending	Stress, strain, and cracking
With carbon black [25,26]	Concrete column embedded with self-sensing concrete into its center	Compression	Strain
	Reinforced concrete beam embedded with self-sensing concrete into uniaxial compression, combined compression and shear, and uniaxial tension zones	Four-point bending	
With hybrid carbon fiber and graphite powder [27]	Concrete column embedded with self-sensing concrete into its center	Compression	Force
With carbon nanotube [28]	Reinforced concrete beam embedded with self-sensing concrete into its tensile region	Three-point bending	Crack propagation and damage accumulation
With carbon fiber or carbon nanofiber [18]	Reinforced concrete beam embedded with self-sensing concrete into its compressive and tensile region	Four-point bending	Strain

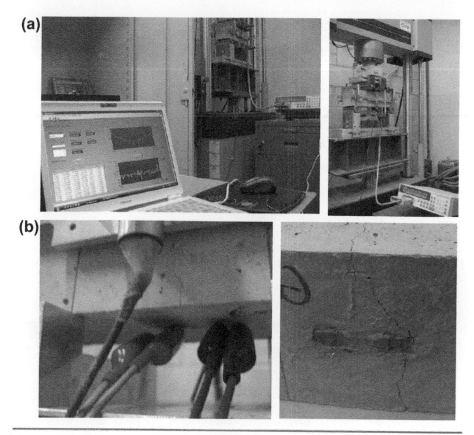

Figure 7.24 Experimental setup [2]. (a) Test setup, (b) specimen.

7.3 Traffic Detection

Vehicle detection is a critical element in traffic management and operations. Currently, various detection systems are being used to collect and process traffic data. These traffic data are mainly obtained from traffic sensors buried under the pavement or installed along the roadway. However, when conventional vehicle detection sensors such as inductive loop, piezoelectric, and optical fiber detectors are buried under the pavement, the pavement life is inevitably be decreased owing to the unfavorable compatibility of sensors with pavement and/or the short lifespan of the sensors [29].

Self-sensing concrete provides a new way to develop vehicle detection sensors. A vehicle detection sensor fabricated with self-sensing concrete has several advantages over conventional detectors, such as easy installation and maintenance, wide detection area, low cost, high anti-jamming ability, long service life, and good

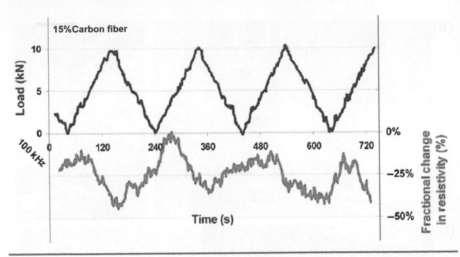

Figure 7.25 Response of self-sensing concrete with carbon fiber to cyclic flexure with 10-kN amplitude [2].

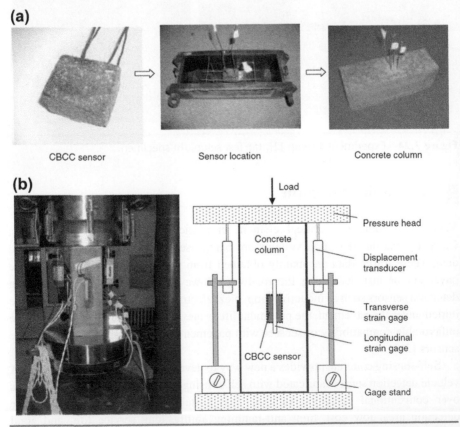

Figure 7.26 Experimental setup [25]. (a) Self-sensing concrete with carbon black (CBCC sensor) and concrete column with embedded CBCC sensor, (b) schematic of uniaxial compressive test setup.

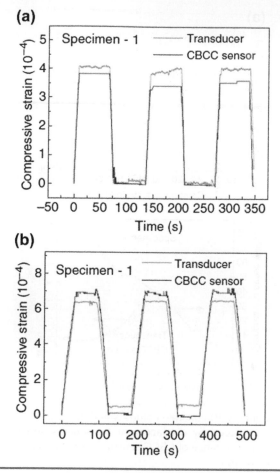

Figure 7.27 Strains of (a) C40 concrete columns and (b) C80 concrete columns under cyclic loading measured by displacement transducers and CBCC (carbon black cement-based composites) sensors [25].

compatibility with pavement structures, because they are made of concrete materials. Pavements or bridge sections embedded with self-sensing concrete, as shown in Figure 7.40, can detect a lot of important traffic data such as traffic flow rates, vehicular speed, and traffic density, and even weighing in motion. Not only can they detect vehicles on roads, they could also be applied in the parking automation system. Moreover, these self-sensing pavements or bridge sections can provide data support for structural health monitoring and condition assessment of traffic infrastructures such as highways and bridges. Previous research on the application of self-sensing concrete for traffic detection is summarized in Table 7.6.

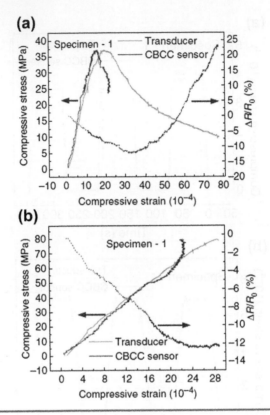

Figure 7.28 Strain of (a) C40 concrete columns and (b) C80 concrete columns under monotonic loading measured by displacement transducers and CBCC sensors, and the fractional change in resistance of CBCC sensors during loading [25]. CBCC, carbon black with concrete column.

Shi and Chung duplicated the weight and motion of a truck traveling on a highway by rotating a car tire between the cylindrical surfaces of two concrete rollers, one of which was made of smart concrete in the lab (as shown in Figure 7.41) [30]. They found that the electrical resistance of smart concrete decreases reversibly with increasing stresses up to 1 MPa and is independent of speed up to 55 miles per hour, which indicates that the self-monitoring concrete containing short carbon fibers (0.5% or 1.0% by weight of cement) is effective for traffic monitoring and weighing in motion (as shown in Figures 7.42 and 7.43) [30].

Wei proposed a monitoring system of vehicle speed based on carbon fiber concrete [31]. The detector was made of a strip component with carbon fiber concrete, and the authors systematically investigated the effect of loading mode and magnitude, electrode layout, testing methods of electrical resistivity, and collection and processing methods of sensing signals on vehicle speed detection

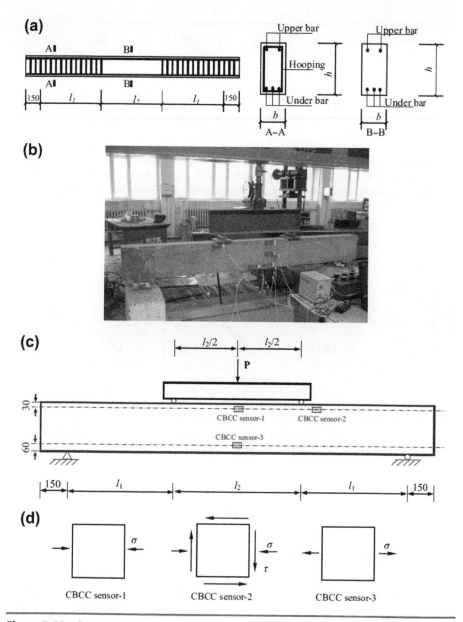

Figure 7.29 Experimental setup [26]. (a) Schematic of concrete beam configurations, (b) experimental setup of a bending concrete beam, (c) schematic of the loading arrangement and CBCC (carbon black cement-based composite) locations, (d) stress states of embedded CBCC sensors.

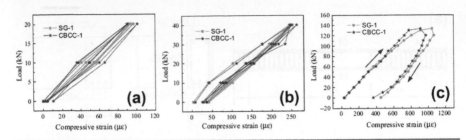

Figure 7.30 Strain sensing property of CBCC sensor-1 (a) under cyclic loading with an amplitude of 20 kN, (b) under cyclic loading with an amplitude of 40 kN, (c) under monotonic loading until te reinforcing bar yielded followed by unloading. CBCC, carbon black with concrete column.

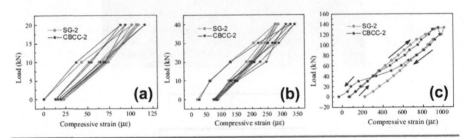

Figure 7.31 Strain sensing property of CBCC sensor-2 (a) under cyclic loading with an amplitude of 20 kN, (b) under cyclic loading with an amplitude of 40 kN, (c) under monotonic loading until the reinforcing bar yielded followed by unloading [26]. CBCC, carbon black with concrete column.

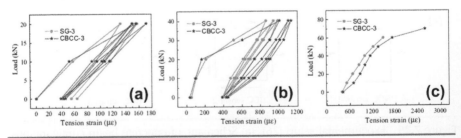

Figure 7.32 Strain sensing property of CBCC sensor-3 (a) under cyclic loading with an amplitude of 20 kN, (b) under cyclic loading with an amplitude of 40 kN, (c) under monotonic loading until the reinforcing bar yielded followed by unloading [26]. CBCC, carbon black with concrete column.

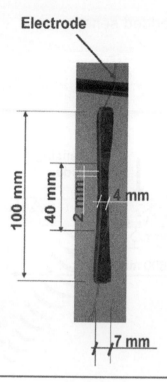

Figure 7.33 Self-sensing concrete with carbon nanotube (embedded SWNTs reinforced cement-based sensors) [28].

through model experiments in the laboratory. Experimental results indicated the feasibility of using carbon fiber concrete to monitor vehicle speed. Jian proposed a weighing-in-motion system based on carbon fiber concrete [32]. She presented the working principle and configuration design scheme of this system, and fabricated strip components with carbon fiber concrete. The effectiveness of weighing in motion was demonstrated in the laboratory for detecting stress from 18 to 100 kN. She also investigated the effect of vehicle speed, loading, and tire contact area with the ground on weighing in motion. Under hierarchical loading, the fractional change in electrical resistance of the concrete strips is also hierarchical according to the level of the load. This result testified to the practicability of applying carbon fiber concrete for weighing in motion. Gong made an improvement to the traffic detection system proposed by Wei and Jian [33]. He used hybrid carbon fiber and carbon black cement mortar with high sensitivity to replace carbon fiber concrete. He also developed programs to calculate vehicle speed, vehicle weight, and traffic flow, and judge vehicle types.

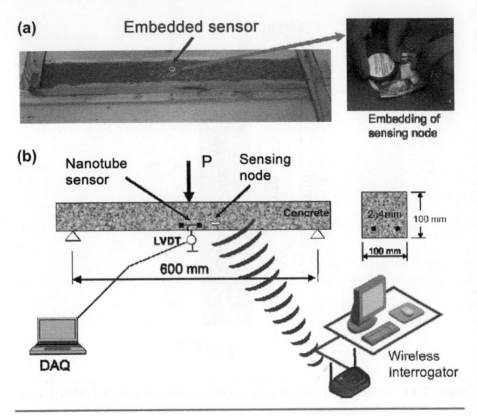

Figure 7.34 Experimental test, (a) embedding of sensing node, (b) wireless monitoring process (DAQ=data acquisition system) [28].

Han et al. developed self-sensing pavement with carbon nanotube cement-based materials and investigated the feasibility of using self-sensing pavement for traffic monitoring with vehicular loading experiments [34]. They found that vehicular loads can lead to remarkable changes in the electrical resistance of carbon nanotube cement-based materials in self-sensing pavements, so traffic flow monitoring and even possible identification of different vehicular loadings could be realized by measuring the electrical resistance of the composites. Han et al. recently extended their previous work on self-sensing carbon nanotube concrete pavements with intensive road tests [35]. They integrated self-sensing carbon nanotube concrete into a controlled pavement test section at the Minnesota Road Research Facility to build a self-sensing concrete pavement system for traffic detection. Road test results showed that the proposed self-sensing pavement system accurately detected the passage of different vehicles under different vehicular speeds and test environments. The developed self-sensing carbon nanotube concrete pavement system could achieve

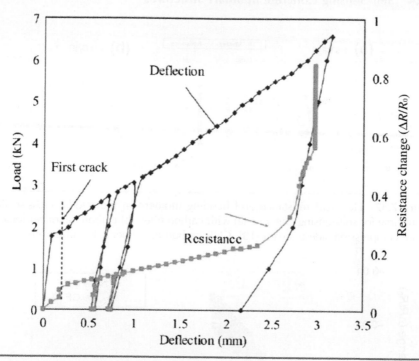

Figure 7.35 Relationship between resistance of self-sensing concrete and load/deflection of beam [28].

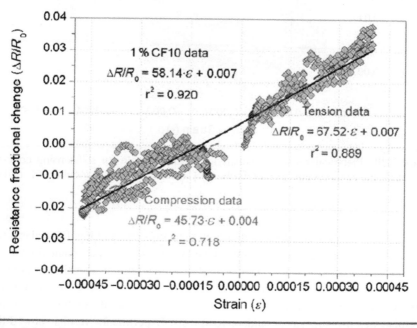

Figure 7.36 Comparison of fractional resistance–strain curves under different strain conditions: compression (left curve) or tension (right curve) [18].

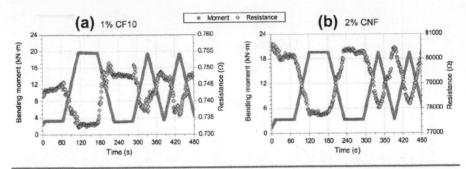

Figure 7.37 Electrical resistance and bending moment in the middle cross-section versus time for self-sensing concrete (a) with carbon fiber, (b) carbon nanofiber located on the compression side of beam [18]. CNF, carbon nanofiber; CF, carbon fiber.

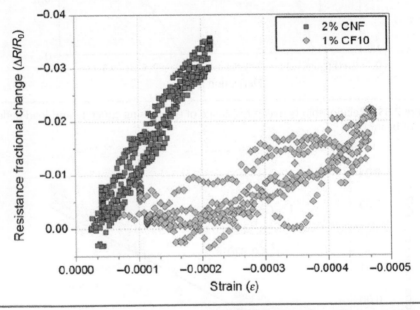

Figure 7.38 Electrical resistance versus longitudinal strain for self-sensing concrete located on the compression side of the beam [18]. CNF, carbon nanofiber; CF, carbon fiber.

real-time vehicle flow detection with a high detection rate and low false-alarm rate [35]. Han et al. also designed another type of self-sensing pavement for vehicle detection [36]. Nickel powder cement-based sensors with high sensitivity were used as vehicle detectors. The sensor arrays were embedded into concrete pavement to form self-sensing pavement. The vehicle detection capability of the self-sensing pavement was investigated in outdoor road tests. Experimental results showed that

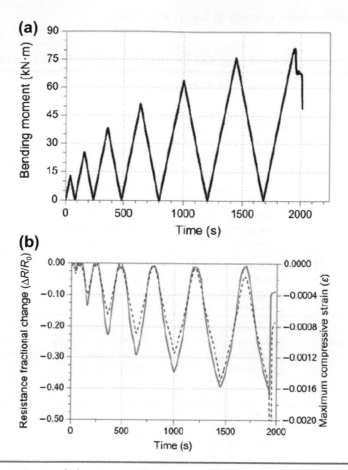

Figure 7.39 Sensing behavior of self-sensing concrete with carbon nanofiber during beam damage test [18]. (a) Bending moment vs time, (b) resistance fractional change (solid curve) and longitudinal strain (dash curve) vs time.

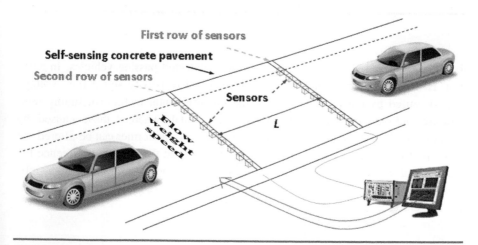

Figure 7.40 Schematic diagram of self-sensing concrete pavement structure for vehicle detection.

TABLE 7.6 Previous Researches on Self-Sensing Concrete for Traffic Detection

Type of Self-Sensing Concrete	Application Style of Self-Sensing Concrete	Test	Detection Goal
With carbon fiber [30]	Self-sensing concrete roller	Rotate a car tire on the roller in lab	Traffic monitoring and weighing in motion
With carbon fiber [31]	Self-sensing concrete strip component integrated into pavement	Test response of self-sensing concrete by using testing machine in lab	Vehicle speed detection
With carbon fiber [32]	Self-sensing concrete strip component integrated into pavement	Test response of self-sensing concrete by using testing machine in lab	Weighting in motion
With hybrid carbon fiber and carbon black [33]	Self-sensing concrete strip component integrated into pavement	Test response of self-sensing concrete by using testing machine in lab	Vehicle speed, vehicle weight and traffic flow detection, and vehicle type judgment
With carbon nanotube [34,35]	Self-sensing concrete strip component integrated into a pavement test section at the Minnesota Road Research Facility	Perform road test at a road research facility with a five-axle semi-trailer truck and a van	Traffic flow monitoring
With nickel powder [36]	Self-sensing concrete arrays integrated into a pavement	Perform road test at outdoor lab with a car	Passing vehicle detection

the proposed self-sensing pavement accurately detected passing vehicles. In addition, Han et al. observed that changes in electrical resistivity signal (i.e., voltage signal) caused by the polarization and the environmental factors (including temperature and humidity) are continuous and gradual, whereas those caused by vehicular loading are transient and abrupt. As a result, the former can be filtered out in the post-processing of measured voltage signals, and they will not influence the accuracy of detection. Therefore, self-sensing pavements embedded with self-sensing concrete features excellent robustness to polarization inside the composites and changes in external environment [36].

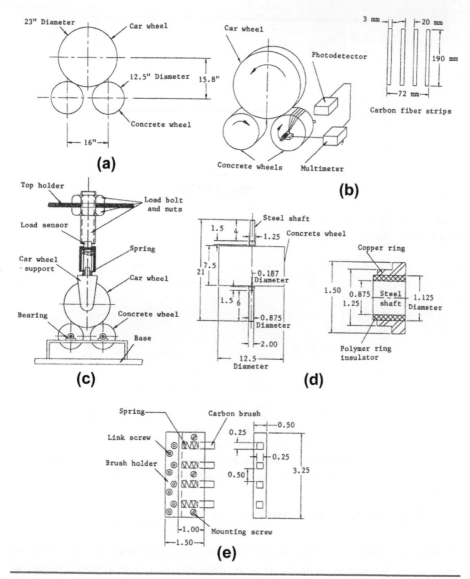

Figure 7.41 Testing setup. (a) Two-dimensional view, (b) three-dimensional view and electrical contact geometry, (c) loading mechanism, (d) steel shaft and copper slip ring, (e) carbon brush assembly [30].

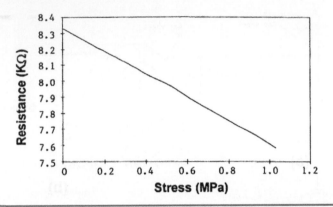

Figure 7.42 Effect of stress on resistance at zero speed for self-monitoring concrete with carbon fibers.

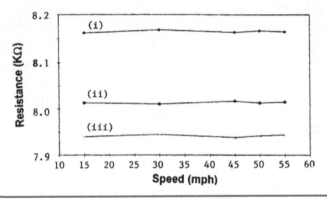

Figure 7.43 Effect of speed on resistance at a constant stress for self-monitoring concrete with carbon fibers (i) 0.21 MPa, (ii) 0.42 MPa, (iii) 0.52 MPa [30].

7.4 Summary and Conclusions

Because self-sensing concrete can work both as a structural material and a sensor in itself, it is a promising candidate to solve durability and incompatibility issues in concrete structures and conventional sensors used for structural health monitoring. Self-sensing concrete can be used in various structural forms such as bulk, coating, sandwich, bonded, and embedded. It can monitor multi-structural parameters such as stress (or force), strain (or deformation), damage, and cracking. Therefore, self-sensing concrete provides a new approach to evaluate the safety and durability of a structure during its service life, to ensure its serviceability and sustainability.

Self-sensing concrete can also be used to develop a vehicle detector. It has several advantages over conventional detectors, such as high detection precision, high anti-jamming ability, easy installation and maintenance, a long service life, and good structural properties. Pavements integrated with self-sensing concrete can detect multi-traffic data such as traffic flow rates, vehicular speed, and traffic density, and even achieve weighing in motion. Therefore, self-sensing concrete provides a new approach to traffic management and control, and health monitoring and condition assessment of infrastructures.

Self-sensing concrete not only has potential in the field of structural health monitoring for concrete structures and traffic detection, it also can be used for military and border security, corrosion monitoring of rebar, structural vibration control, and so on. It can ensure the safety, durability, serviceability, and sustainability of infrastructures such as high-rise buildings, large-span bridges, tunnels, high-speed railways, offshore structures, dams, and nuclear power plants.

References

[1] Ou JP, Li H. Structural health monitoring in mainland China: review and future trends. Struct Health Monit 2010;9(3):219–31.

[2] Azhari F. Cement-based sensors for structural health monitoring [dissertation for the Master Degree of Applied Science]. Canada: University of British Columbia; 2008.

[3] Sett K. Characterization and modeling of structural and self-monitoring behavior of fiber reinforced polymer concrete [dissertation for the Master of Science in Civil Engineering]. USA: University of Houston; 2003.

[4] Zhang DX, Luo ZH, Luo ZP, Wu SG. Sensitivity of reinforced components of CFRC. J Harbin Inst Technol 2004;36(10):1411–3.

[5] Zhang DX, Wu SG, Ma B, Zhao JH. Sensitivities of carbon fiber reinforced concrete under bending loading. J Jilin Univ Eng Technol Ed 2004;34(4):679–83.

[6] Zhang W, Xie HC, Liu JW, Shi B. Experimental study on elastic stress self-monitoring of carbon fiber reinforced smart concrete beams. J Southeast Univ Nat Sci Ed 2004;34(5):647–50.

[7] Hou TC, Lynch JP. Conductivity-based strain monitoring and damage characterization of fiber reinforced cementitious structural components. Proc SPIE 2005;5765:419–29.

[8] Wang W, Wu SG, Dai HZ. Fatigue behavior and life prediction of carbon fiber reinforced concrete under cyclic flexural loading. Mater Sci Eng A 2006;434:347–51.

[9] Zhang DJ, Xu SL, Hao HM. Experimental study on fracture parameters of three-point bending beam based on smart properties of CFRC. J Hydroelectr Eng 2008;27(2):71–7.

[10] Howser RN, Dhonde HB, Mo YL. Self-sensing of carbon nanofiber concrete columns subjected to reversed cyclic loading. Smart Mater Struct 2011;20:085031 (13pp).

[11] Saafi M, Andrew K, Tang PL, McGhon D, Taylor S, Rahman M, et al. Multifunctional properties of carbon nanotube/fly ash geopolymeric nanocomposites. Constr Build Mater 2013;49:46–55.

[12] Prashanth P, Vipulanandan C. Characterization of thin disk piezoresistive smart material for hurricane applications. THC-IT 2009 Conference and Exhibition; 2009. 1–2.

[13] Wen SH, Chung DDL. Carbon fiber-reinforced cement as a strain-sensing coating. Cem Concr Res 2001;31(4):665–7.

[14] Huang LN, Zhang DX, Wu SG, Zhao JH. Study on pulling sensitivity character of CFRC and smart monitoring of beam specimens. J Mater Eng 2005;2:26–9, 33.

[15] Zhang DX, Su YQ, Huang LN, Wu SG, Zhao JH. Experimental research on the smart character of laminated CFRC bending test specimens. Low Temp Archit Technol 2005;103(1):62–4.

[16] Hong L. Study on the smart properties of the graphite slurry infiltrated steel fiber concrete [dissertation for the Doctor Degree in Engineering]. China: Dalian University of Technology; 2006.

[17] Wang W, Dai HZ, Wu SG. Mechanical behavior and electrical property of CFRC-strengthened RC beams under fatigue and monotonic loading. Mater Sci Eng A 2008;479:191–6.

[18] Baeza FJ, Galao O, Zornoza E, Garcés P. Multifunctional cement composites strain and damage sensors applied on reinforced concrete (RC) structural elements. Materials 2013;6:841–55.

[19] Zheng LX, Song XH, Li ZQ. Self-monitoring of the deformation in smart concrete structures. J Huazhong Univ Sci Technol 2004;32(4):30–1, 55.

[20] Hong L, Wang SM. Study on the stress-resistance effect of the graphite slurry infiltrated steel fiber concrete. Funct Mater; 2010:135–7.

[21] Wu SG, Dai HZ, Wang W. Effect of CFRC layers on the electrical properties and failure mode of RC beams strengthened with CFRC composites. Smart Mater Struct 2007;16:2056–62.

[22] Chen B, Liu JY. Damage in carbon fiber-reinforced concrete, monitored by both electrical resistance measurement and acoustic emission analysis. Constr Build Mater 2008;22:2196–201.

[23] Han BG. Properties, sensors and structures of pressure-sensitive carbon fiber cement paste [dissertation for the Doctor Degree in Engineering]. China: Harbin Institute of Technology; 2005.

[24] Chacko RM. Carbon-fiber reinforced cement based sensors [dissertation for the Master Degree of Applied Science]. Canada: University of British Columbia; 2005.

[25] Xiao HG, Li H, Ou JP. Self-monitoring properties of concrete columns with embedded cement-based strain sensors. J Intell Mater Syst Struct 2011;22(2):191–200.

[26] Xiao HG, Li H, Ou JP. Strain sensing properties of cement-based sensors embedded at various stress zones in a bending concrete beam. Sens Actuators Phys 2011;167(2):581–7.

[27] Fan XM, Ao F, Sun MQ, Li ZQ. Piezoresistivity of carbon fiber graphite cement-based composites embedded in concrete column. J Build Mater 2011;14(1):88–91.

[28] Saafi M. Wireless and embedded carbon nanotube networks for damage detection in concrete structures. Nanotechnology 2009;20. 395502 (7pp).

[29] Klein LA, Mills MK, Gibson RP. In: Traffic detector handbook. 3rd ed. U.S. Department of Transportation, Federal Highway Administration; 2006.

[30] Shi ZQ, Chung DDL. Carbon fiber-reinforced concrete for traffic monitoring and weighing in motion. Cem Concr Res 1999;29:435–9.

[31] Wei WB. A research of the traffic vehicle-speed measuring system based on the pressure-sensitivity of CFRC [dissertation for the Master Degree in Engineering]. China: Shantou University; 2003.

[32] Jian HL. Research of the traffic weighting monitoring system based on the pressure-sensitivity of CFRC [dissertation for the Master Degree in Engineering]. China: Shantou University; 2004.

[33] Gong ZQ. Research of the traffic monitoring system based on the pressure-sensitivity of CFRM [dissertation for the Master Degree in Engineering]. China: Shantou University; 2007.

[34] Han BG, Yu X, Kwon E. A self-sensing carbon nanotube/cement composite for traffic monitoring. Nanotechnology 2009;20. 445501 (5pp).

[35] Han BG, Zhang K, Burnham T, Kwon E, Yu X. Integration and road tests of a self-sensing CNT concrete pavement system for traffic detection. Smart Mater Struct 2013;22:015020 (8pp).

[36] Han BG, Zhang K, Yu X, Kwon E, Ou JP. Nickel powder-based self-sensing pavement for vehicle detection. Measurement 2011;44(9):1645–50.

Chapter **8**

Carbon-Fiber-Based Self-Sensing Concrete

Chapter Outline

231

Self-Sensing Concrete in Smart Structures. http://dx.doi.org/10.1016/B978-0-12-800517-0.00008-3

8.1 Introduction and Synopsis

Carbon fiber, one type of graphite, consists of fibers of about 5–10 μm in diameter and is mostly composed of carbon atoms. The carbon atoms are bonded together in crystals that are more or less aligned parallel to the long axis of the fiber (as shown in Figure 8.1(a)) [1,2]). Depending on precursor fiber materials, carbon fiber can be classified into four types, namely polyacrylonitrile (PAN)-based carbon fiber, pitch-based carbon fiber, rayon-based carbon fiber, and gas-phase-grown carbon fiber [5]. Carbon fiber has been extensively applied in fabricating smart and multifunctional composites due to its excellent physical and chemical properties. It is inert in aggressive environments, abrasion-resistant and stable at high temperature, medically safe, as strong as steel fiber, and more chemically stable than glass fiber in alkaline environments. Moreover, carbon fiber is low in density, especially when compared with steel fiber, and its strength to density ratio is one of the highest among all fiber types. In addition, carbon fiber possesses an advantage of high electrical conductivity [6]. When used as functional fillers of the self-sensing concrete, carbon fiber is usually in the form of chopped carbon fiber (as shown in Figure 8.1(b) [3,4]. The chopped carbon fiber represents a high aspect ratio due to its millimeter-scale length and microscale diameter. Chen and Chung first successfully fabricated carbon fiber cement mortar with smart properties by using short-cut isotropic pitch-based carbon fiber with a length of 5.1 mm and a diameter of 10 μm in 1993 [7]. Mao et al.

(a)

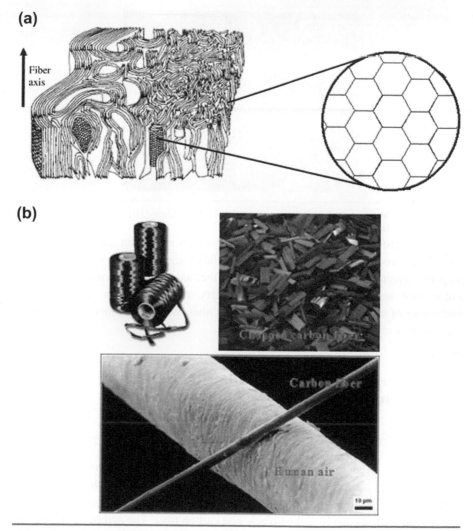

(b)

Figure 8.1 Carbon fiber. (a) Structure of carbon fiber [1,2], (b) carbon fiber and short-cut carbon fiber [3,4].

subsequently fabricated carbon fiber cement paste with pressure-sensitive properties by using PAN-based carbon fiber with a length of 5 mm and a diameter of 7 µm in 1996 [8]. Azhari employed petroleum pitch-based carbon fiber with a length of 6 mm and a diameter of 18 µm and coal-tar pitch-based carbon fiber with a length of 6 mm and a diameter of 11 µm to develop pressure-sensitive concrete [9]. Up to now, the self-sensing concrete with carbon fiber (an example is shown in Figure 8.2) is the most extensively and comprehensively studied among all types of self-sensing concrete.

Figure 8.2 Intrinsic self-sensing concrete with carbon fiber [10]].

This chapter will introduce the authors' research results on carbon-fiber-based self-sensing concrete, with attentions to the fabrication, measurement of sensing property, sensing property, and improvement methods. The performance, effect of temperature and humidity on sensing property, and application in stress/strain monitoring of concrete components of carbon fiber concrete sensors are also introduced in this chapter [11–15].

8.2 Fabrication of Carbon-Fiber-Based Self-Sensing Concrete

8.2.1 RAW MATERIALS

Raw materials used to produce self-sensing concrete with carbon fiber mainly include Portland cement, silica fume, dispersant agent (methylcellulose), defoamer agent (tributyl phosphate), water-reducing agent (sodium salt of a condensed naphthalene sulfonic acid), carbon black, and carbon fiber. The carbon fiber used for this study is PAN-based carbon fiber made by Jilin Carbon Co. Ltd, and its properties are shown in Table 8.1.

The electrode material and conductive paint are selected based on the following fundamental principles: (1) The electrode material and conductive paint should possess excellent electrical conductivity to minimize the effect of electrode

TABLE 8.1 Properties of Carbon Fiber

Filament Diameter (μm)	Tensile Strength (MPa)	Tensile Modulus (GPa)	Specific Gravity (g/cm³)	Resistivity (Ω cm)
7	3450	230	1.8	3×10^{-3}

TABLE 8.2 Electrode Material Properties and Electrode Fixing Styles

Material	Size (cm × cm)	Thickness (mm)	Resistance (Ω)	Fixing Style
Copper foil	4 × 5	0.3	0.106	Pasted
Copper gauze	4 × 5	0.5	0.125	Embedded

resistance and contact resistance. (2) The electrode material should have little or no, if possible, effect on the mechanical and electrical properties of concrete with carbon fiber. (3) The electrode material should be resistant to corrosion in concrete with carbon fiber and have good durability to meet the requirement of long-term monitoring. Meanwhile, conductive paint should be resistant to aging. Based on these principles, the electrode material and conductive paint used in the following experiments are copper foil, copper gauze, and silver paint respectively. The material properties and the fixing styles are shown in Table 8.2 and Table 8.3.

8.2.2 FABRICATION METHOD

Carbon fiber, dispersant agent (0.4% by weight of cement), water-reducing agent (1.5% by weight of cement), and water were mixed and stirred in a mortar mixer for about 3 min, and then carbon black (for concrete with hybrid carbon fiber and carbon black), silica fume (15% by weight of cement), cement, and defoamer agent (0.05% by weight of cement) were added and mixed for another 3 min. After pouring the mixture into a number of oiled molds and embedding the gauze electrode, a vibrator was used to remove air bubbles. The specimens were removed from the molds in 24 h, and cured then in a standard fog room at 20 °C and 100% relative humidity for 28 days. The specimens were dried for a period of time before they are tested at ambient temperature.

8.3 Measurement of Sensing Property of Carbon-Fiber-Based Self-Sensing Concrete

The sizes of specimens are $4 \times 4 \times 4$ cm, $4 \times 4 \times 5$ cm, $4 \times 4 \times 13$ cm, and $4 \times 4 \times 17$ cm, respectively. Two electrode fixing styles are shown in Figure 8.3. The

TABLE 8.3 Properties of Silver Paint

Composition		Resistivity (Ω cm)	Shear Strength (MPa)
Silver	Epoxy resin	$<2 \times 10^{-3}$	>12

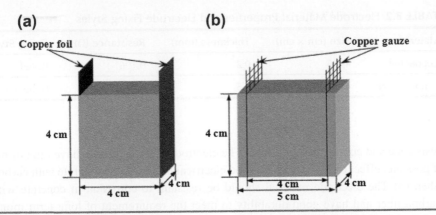

Figure 8.3 Schematic diagram of two electrode fixing styles: (a) pasted electrode; (b) embedded electrode [13].

pasted electrode means the two copper foils were pasted by silver paint on the two parallel planes of specimens, whereas the embedded electrode means the embedment of poles into the two parallel planes of specimens.

8.3.1 COMPARISON OF TWO ELECTRODE FIXING STYLES

To compare the two electrode fixing styles shown in Figure 8.3, the resistance of specimens (3.00 vol. % carbon fiber) with different electrode fixing styles is measured. The mean values and coefficients of variability of the resistance of specimens are listed in Table 8.4. The coefficient of variability is calculated by

$$C_v = \frac{\sqrt{\sum_{i=1}^{n} (R_i - \overline{R})^2 / (n-1)}}{\overline{R}} \tag{8.1}$$

where R_i (i:1–6) is the resistance of specimens and $\overline{R}$ is the mean resistance of specimens.

The mean values and coefficients of variability of the resistance of specimens with pasted copper foil electrode are much larger than those of specimens with embedded copper gauze electrode, as listed in Table 8.4.

TABLE 8.4 Mean Values and Coefficients of Variability of the Resistance of Specimens

Fixing Styles of Electrode	Embedded Copper Gauze	Pasted Copper Foil
$\overline{R}$ (Ω)	2.38	3.51
C_v (%)	8.70	28.70

In fact, the resistance of specimens mainly consists of the true resistance of concrete with carbon fiber, the resistance of electrode, and the contact resistance between electrode and concrete with carbon fiber. Among these three kinds of resistance, the resistance of electrode can be negligible compared with the true resistance of concrete with carbon fiber. Therefore the difference between the true resistance and the measured resistance mainly results from the contact resistance. The larger the measured resistance, the greater the contact resistance. The higher coefficients of variability calculated for specimens with pasted electrode also indicate that the scatter of measured resistance is caused by the contact resistance.

From Figure 8.4, it can be seen that the fractional change in resistivity of carbon fiber concrete with embedded electrode is relatively smooth, but that of carbon fiber concrete with pasted electrode is relatively high and presents a sharp drop at the initial stage of loading, which is attributed to the poor bond between copper foil and carbon fiber concrete. Under compression, the contact resistance diminishes because the bond between copper foil and carbon fiber concrete improves.

From these results and analysis, it can be concluded that the copper gauze electrode is suitable for concrete with carbon fiber, and the embedded style should be used to fix the copper gauze electrode.

8.3.2 CHARACTERISTICS OF EMBEDDED GAUZE ELECTRODE

8.3.2.1 Bond between Gauze Electrode and Carbon Fiber Concrete

The compressive strength of specimens with embedded gauze electrode which has different mesh sizes is listed in Table 8.5. It is shown that the compressive strength of specimens with gauze electrode (mesh size of which is 1.66×1.66 mm) is relatively

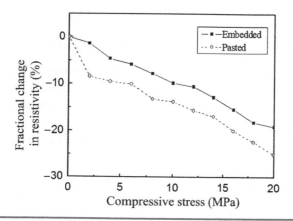

Figure 8.4 Compressive stress and fractional change in resistivity of concrete with carbon fiber during static loading [13].

TABLE 8.5 Compressive Strength of Specimens without Electrode and with Gauze Electrode Having Different Mesh Sizes

Measured Specimens	Compressive Strength (MPa)	Coefficient of Variability (%)
Without electrode	56.6	2.76
With gauze electrode (mesh size of which is 1.66 × 1.66 mm)	50.8	4.71
With gauze electrode (mesh size of which is 2.03 × 2.03 mm)	57.3	3.05
With gauze electrode (mesh size of which is 2.56 × 2.56 mm)	55.9	2.81

low and has higher scatter than specimens without electrode, but the compressive strength of specimens with gauze electrode (mesh sizes of which are 2.03 × 2.03 mm and 2.56 × 2.56 mm) is unaffected by the embedded electrode.

In addition, the observation of the failure modes shows that the bond between electrode and carbon fiber concrete is weak for specimens with gauze electrode having mesh size of 1.66 × 1.66 mm, because most of them fail at the interface between electrode and carbon fiber concrete. Whereas no interfacial failure is observed for the specimens with gauze electrode having mesh sizes of 2.03 × 2.03 mm and 2.56 × 2.56 mm.

Through adjusting the mesh size of gauze electrode, the bond between electrode and carbon fiber concrete can be improved, which results in the little influence of gauze electrode on the compressive strength of carbon fiber concrete and good compatibility of gauze electrode with carbon fiber concrete.

8.3.2.2 Ohmic Characteristics of Gauze Electrode

It is well known that excellent electrode should be the ohmic contact electrode with the following characteristics: (1) No rectification characteristic. The resistance is independent of the electric field direction. (2) Voltage–current characteristic should be linear at the low voltage. (3) Authenticity. Contact resistance is small enough to negligible compared with the true resistance of measured objects [16].

The relationship between current and voltage of carbon fiber concrete with embedded gauze electrode, as shown in Figure 8.5, coincides well with characteristic (1) and characteristic (2). In addition, the resistance of carbon fiber cement paste listed in Table 8.6 meets the series relationship, which indicates that the measured resistance is true and coincides with characteristic (3). Otherwise, the resistance summation of B and C would be higher than the resistance of $B + C$.

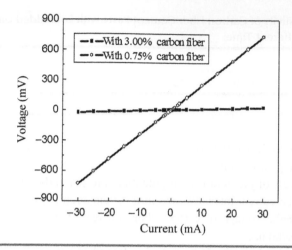

Figure 8.5 Current and voltage of carbon fiber concrete [13].

8.3.2.3 Durability of Gauze Electrode

The resistivity of carbon fiber concrete with embedded gauze electrode measured at different time (without loading) is listed in Table 8.7. The resistivity of carbon fiber cement paste with 3.00 vol. % carbon fiber hardly has change during one year. Moreover, the electrode in the crushed specimen is not corroded. It is shown that the gauze electrode has good durability.

8.3.2.4 Polarization Comparison of Gauze Electrode and Foil Electrode

There exist a large number of ions, such as K^+, Na^+, Ca^+, OH^-, and SO_4^{2-}, etc., in carbon fiber concrete, Under the external voltage, the ions would move opposite to the external electric field. Thus the positive ions and the negative ions will gather at

TABLE 8.6 Resistance of Specimens with Two Fiber Fractions Measured by Different Methods

Measuring Method	Volume Fraction of Carbon Fiber 3%			Volume Fraction of Carbon Fiber 0.75%		
	B	*C*	*B + C*	*B*	*C*	*B + C*
Direct-current two-pole method (Ω)	2.37	2.42	3.15	77.30	80.10	99.40
Direct-current four-pole method (Ω)	0.86	0.84	1.90	24.60	26.20	50.80

TABLE 8.7 Resistivity of Carbon Fiber Cement Paste with Embedded Gauze Electrode
Measured at Different Time

Time/mo.	2	3	6	8	10	12
Resistivity (Ω cm)	9.50	9.53	9.49	9.50	9.48	9.51

the negative electrode and the positive electrode, respectively. These bound charges
will generate a polarization potential opposite to the external voltage [17]. The
measuring process of the polarization potential is to put carbon fiber concrete in a
circuit, adjust the current intensity, then turn off the circuit after half a minute, and
finally measure the voltage of carbon fiber concrete. The measured value is the
polarization potential.

The measured polarization potential of carbon fiber concrete (0.75 vol.% carbon
fiber) with embedded gauze electrode and pasted foil electrode at different current is
shown in Figure 8.6. It shows that the polarization of specimens with gauze electrode
is weaker than that of specimens with foil electrode, and the difference between them
increases with current.

The polarization is equivalent to a charge process of condenser. The relationship
between the capacitance and the configuration parameters of capacitor is given by

$$C = \frac{\varepsilon S}{l} \tag{8.2}$$

where C is the capacitance, S is the polar plate area, l is the space between polar
plates, and ε is the dielectric constant.

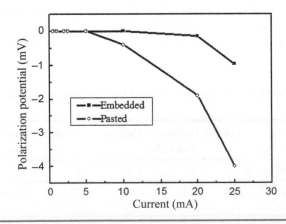

Figure 8.6 Polarization potential of specimens with various electrode fixing styles
measured at different current [13].

From Eq. (8.2), the capacitance of specimens with gauze electrode is smaller than that of specimens with foil electrode when the space between polar plates is same because the former has a smaller effective area. Therefore, the polarization potential of specimens with gauze electrode is smaller because the gauze electrode gathers less charge under the same external voltage, which is obtained by

$$Q = CU \tag{8.3}$$

where U is the external voltage.

8.3.3 COMPARISON OF TWO ELECTRODE LAYOUTS

8.3.3.1 Testing Principles of Two Electrode Layouts

The electrode of carbon fiber concrete is commonly set in the two-pole layout and the four-pole layout. The embedded gauze electrode can also be set in these two layouts as shown in Figure 8.7. The two-pole layout is embedding two poles in the carbon fiber cement paste and the corresponding measuring method is the direct-current two-pole method. The four-pole layout is embedding four poles in carbon fiber cement paste and the corresponding measuring method is the direct-current four-pole method. In the direct-current two-pole method, the two poles are both current electrode and voltage electrode, whereas the inner two poles and the outer two poles are voltage electrode and current electrode, respectively, in the direct-current four-pole method.

8.3.3.2 Influence of Contact Resistance on Testing Accuracy

To compare the direct-current two-pole method with the direct-current four-pole method, the resistance values of two kinds of specimens with 3.00 vol.% carbon fiber

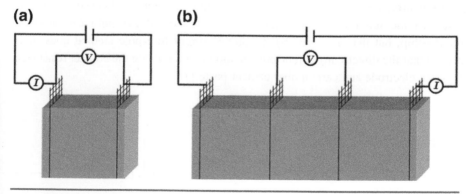

Figure 8.7 Sketch maps of the direct-current two-pole method and the direct-current four-pole method: (a) Two-pole layout; (b) Four-pole layout [13].

TABLE 8.8 Resistance of Specimens Measured by Different Methods

Measuring Method	Direct-Current Two-Pole Method	Direct-Current Four-Pole Method
Resistance (Ω)	2.38	0.87

measured by these two methods are listed in Table 8.8. It shows that the resistance measured by the direct-current four-pole method is less than that obtained by the direct-current two-pole method. The equivalent circuit diagram is shown in Figure 8.8. The voltage measured by the direct-current two-pole method (U) includes the voltage of the electrode (U_E), the voltage of carbon fiber cement paste (U_R) and the voltage of contact resistance (U_C). Because U_E is small enough to be negligible, the voltage measured by the direct-current two-pole method mainly includes U_R and U_C Whereas the voltage measured by the direct-current four-electrode method is only U_R.

8.3.3.3 Calculation of Contact Resistance

Two kinds of carbon fiber concrete with 3.00 vol. % and 0.75 vol.% carbon fiber, respectively, are prepared to analyze the effect of contact resistance on the testing accuracy of resistance. The electrode layout is shown in Figure 8.9.

The resistance of specimen B, C, and $B + C$ in Figure 8.9, which is measured by the direct-current two-pole method and the direct-current four-pole method, respectively, is listed in Table 8.9. For carbon fiber cement paste with two fiber fractions, the resistance measured by the direct-current four-pole method is smaller than that measured by the direct-current two-pole method, and the more the carbon fiber fraction, the larger the effect of contact resistance.

Comparing the resistance summation of B and C with the resistance of $B + C$, the resistance measured by the direct-current two-pole method does not meet the series relationship, but that measured by the direct-current four-pole method does. It is proved that the direct-current four-pole method can eliminate the contact resistance between electrode and carbon fiber cement paste [18].

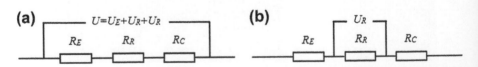

Figure 8.8 Equivalent circuit diagram for (a) direct-current two-pole method and (b) direct-current four-pole method [13].

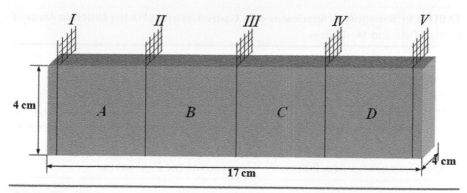

Figure 8.9 Sketch map of the specimen used to calculate contact resistance [13].

The contact resistance of pole *II*, *III*, *IV*, and carbon fiber cement paste is calculated by Equation set Eq. (8.6) and the results are listed in Table 8.9. The contact resistance takes greater proportion of the resistance measured by the direct-current two-pole method.

$$
\begin{cases}
R_{II} + R_{4B} + R_{4C} + R_{IV} = R_{2(B+C)} \\
R_{II} + R_{4B} + R_{III} = R_{2B} \\
R_{III} + R_{4C} + R_{IV} = R_{2C} \\
R_{4B} + R_{4C} = R_{4(B+C)}
\end{cases}
\tag{8.4}
$$

where R_{2B}, R_{2C}, and $R_{2(B+C)}$ are the resistances of specimens *B*, *C*, and *B* + *C* measured by the direct-current two-pole method, respectively. R_{4B}, R_{4C}, and $R_{4(B+C)}$ are the resistances of specimens *B*, *C*, and *B* + *C* measured by the direct-current four-pole method, respectively. R_{II}, R_{III}, and R_{IV} are the contact resistance of pole *II*, *III*, and *IV*, and carbon fiber concrete respectively.

This analysis indicates the contact resistance between the electrode and carbon fiber concrete causes the difference between the resistance measured by these two methods, and the direct-current four-pole method can eliminate it. Therefore the four-pole layout is suitable for setting the embedded gauze electrode and the direct-current four-pole method is suitable for measuring the resistance of concrete with carbon fiber.

TABLE 8.9 Contact Resistance of Poles and Carbon Fiber Concrete

Contact Resistance	R_{II}	R_{III}	R_{IV}	$R_{II} + R_{III}$	$R_{II} + R_{IV}$	$R_{III} + R_{IV}$
Specimen with 3.00 vol.% carbon fiber (Ω)	0.59	0.92	0.66	1.51	1.25	1.58
Specimen with 0.75 vol.% carbon fiber (Ω)	23.70	29.00	24.90	52.70	48.60	53.90

TABLE 8.10 Resistivity of Specimens with Gauze Electrode Having Different Areas of Voltage Pole and Mesh Sizes

Mesh Size of Gauze Electrode	2.56 × 2.56 mm			2.03 × 2.03 mm		
Area fraction of pole to the section of specimen (%)	30	60	100	30	60	100
Resistivity (Ω cm)	3.44	3.48	3.41	3.36	3.40	3.32

8.3.4 Optimization of Electrode Configuration Parameters

8.3.4.1 Determination of Voltage Electrode Geometric Parameters

The influence of the area of voltage pole and the mesh size of gauze electrode on the resistivity of the specimens with 3.00 vol. % carbon fiber is listed in Table 8.10. The results show the resistivity does not change with them.

The resistivity of carbon fiber concrete is calculated by

$$\rho = \frac{RS}{L} \tag{8.5}$$

where S is the effective area of voltage pole and the effective area of the mesh, L is the space between two voltage poles, and R is the resistance of carbon fiber concrete that is given by

$$R = V \bigg/ \left(\frac{I}{S_0}S\right) = \frac{VS_0}{IS} \tag{8.6}$$

where S_0 is the sectional area of specimen, V is the testing voltage, and I is the testing current.

Equation (8.7) can be derived from Eqs (8.3) and (8.4).

$$\rho = \frac{VS_0}{IL} \tag{8.7}$$

It shows that the resistivity is independent of the area of voltage pole and the mesh size of gauze electrode. Therefore, the area of voltage pole can be properly decreased and the mesh size of gauze electrode can be properly increased in order to reduce the influence of gauze electrode on the mechanical properties of carbon fiber concrete.

8.3.4.2 Determination of the Space between Current Pole and Voltage Pole

From Table 8.11, it can be seen that the space between current pole and voltage pole influences the resistivity of carbon fiber concrete when less than 0.75 cm, but does

TABLE 8.11 Resistivity Measured at Different Spaces between Current Pole and Voltage Pole

Space between Current Pole and Voltage Pole (cm)	0.25	0.50	0.75	1.00	1.50	2.00	4.00	8.00
Resistivity (Ω cm)	5.16	4.16	3.44	3.44	3.43	3.44	3.42	3.44

not influence the resistivity when more than 0.75 cm. This indicates that the nonuniform electrical field close to current pole has an effect on the measured voltage when the space between current pole and voltage pole is small.

8.3.5 DESIGN OF DATA ACQUISITION SYSTEM

8.3.5.1 Principle of Data Acquisition System

The sensing mechanism of self-sensing concrete with carbon fiber is monitoring their own strain, stress, and damage through the change in resistance, so the monitoring signal of this kind of sensor is resistance signal. In the direct-current four-pole method, which needs synchronously collecting current and voltage signals, the resistance is calculated by voltage and current. The voltage signal can be collected by A/D card, but the current signal cannot. A reference resistance with constant value and a carbon fiber concrete are applied to constitute a series module as shown in Figure 8.10.

The current in the series module changes with time and can be described as

$$I(t) = \frac{U_c(t)}{R_c(t)} = \frac{U_r(t)}{R_r} \tag{8.8}$$

where $R_c(t)$ and $U_c(t)$ are the resistance and the voltage of self-sensing concrete with carbon fiber, respectively. R_r and $U_r(t)$ are the resistance and the voltage of reference resistance, respectively.

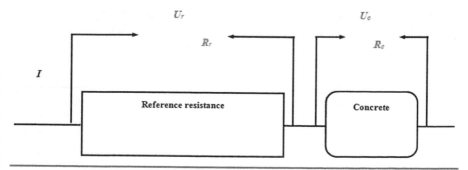

Figure 8.10 Series module of carbon fiber concrete and reference resistance [13].

Equation (8.8) can be further derived as

$$R_c(t) = \frac{U_c(t)}{U_r(t)} R_r \qquad (8.9)$$

And the fractional change in resistivity can be calculated by

$$\Delta R_c(t) = \left(\frac{U_c(t)}{U_r(t)} - \frac{U_c(t_0)}{U_r(t_0)} \right) R_r \qquad (8.10)$$

where $U_r(t_0)$ is the initial voltage of reference resistance and $U_c(t_0)$ is the initial voltage of carbon fiber concrete.

In Eq. (8.11), $U_c(t)$ and $U_r(t)$ are synchronously collected by an A/D card. Thus the fractional change in resistivity of carbon fiber concrete can be derived by the voltage signal $U_c(t)$ and $U_r(t)$.

8.3.5.2 Working Process of Data Acquisition System

The working process of the data acquisition system is shown in Figure 8.11. Two analog voltage signals of carbon fiber concrete and reference resistance are transferred to the A/D card and transformed into digital signals after being treated by the signal collection and amplifier of the front-end processing circuit, and they are then transmitted to and processed in a computer or mono-plate processor.

8.3.5.3 Circuit Design of Multichannel Data Acquisition System

Many carbon fiber concrete sensors are needed to realize health monitoring of concrete structures. They can be arranged in a circuit as shown in Figure 8.12. In this way, the number of power sources can be reduced and the sensors do not interfere with each other, which results in relatively high reliability and real-time multichannel acquisition of the data acquisition system.

The data acquisition system developed according to this principle is shown in Figure 8.13. From the principle and working process of the data acquisition system, the data acquisition system developed has characteristics such as simple circuit, good practicability, high precision, and ability to realize real-time on-line acquisition.

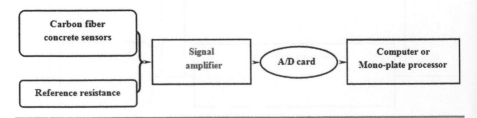

Figure 8.11 Working process of data acquisition system [13].

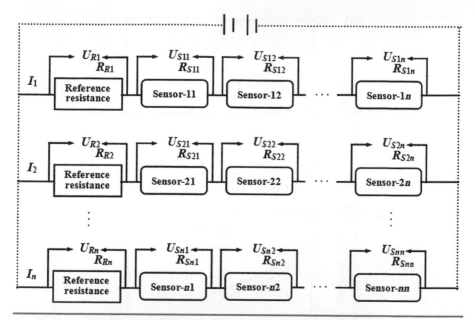

Figure 8.12 Circuit diagram of multichannel data acquisition [13].

Figure 8.13 Data acquisition system [13].

8.4 Sensing Property and Its Improvement of Carbon-Fiber-Based Self-Sensing Concrete

The repeatability of sensing property is the prerequisite condition for using carbon fiber concrete to produce sensors [19,20]. It can be seen from Figure 8.14 that the

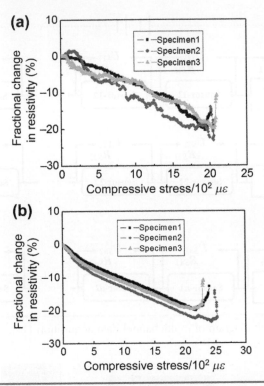

Figure 8.14 Fractional changes in resistivity of concrete with carbon fiber only and with hybrid carbon fiber and carbon black under compressive load [14]. (a) With carbon fiber only, (b) with hybrid carbon fiber and carbon black.

sensing property of concrete with hybrid carbon fiber and carbon black has better repeatability and linearity than that of concrete with carbon fiber only. Sensing property results from the variation in the contact resistance among fillers and between fillers and concrete matrix under compressive loading, which is due to the contacting and tunneling conduction effect [21,22]. The combination of carbon fibers and carbon black particles provides charge transport over long and short distances and enhances the contacting and tunneling conduction effect as shown in Figure 8.15 [23,24], and consequently keeps the response of contact resistance to a force-field sensitive and stable. Therefore, the repeatability and the linearity of sensing property of concrete composites can be improved by adding carbon fiber and carbon black into concrete.

It can be seen from Figure 8.16(a) that the fractional change in electrical resistivity decreases as the compressive stress/strain increases and can go up to 27% maximum, and it can be seen from Figure 8.16(b) that the compressive stress

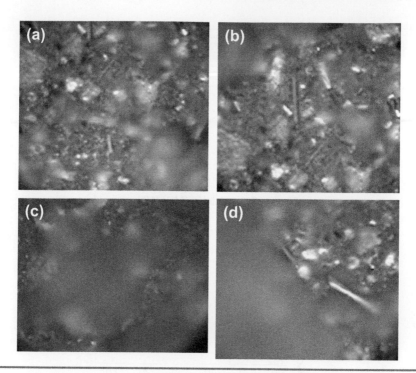

Figure 8.15   Microscopical images for concrete with hybrid carbon fiber and carbon black [11]. (a) General pattern, (b) fiber contact, (c) carbon chains, (d) bridge connection of fiber and black carbon.

approximately linearly increases with the compressive strain until the specimens are crushed, and the ultimate stress is about 40 MPa.

As shown in Figures 8.17 and 8.18, the relationships between the fractional change in electrical resistivity and the compressive stress/strain of concrete with hybrid carbon fiber and carbon black have different change trends under the repeated compressive loads with amplitudes of 12 and 24 kN. Figure 8.17 illustrates that the fractional change in electrical resistivity decreases reversibly upon loading and increases reversibly upon unloading in every cycle. However, it can be seen from Figure 8.18 that the fractional change in electrical resistivity is irreversible upon loading and unloading in every cycle because its original value decreases with the loading and unloading cycle.

The sensing property of concrete results from the variation in the contact resistance among the electrical fillers and between the electrical fillers and concrete under compressive loading, as a result of the deformation of concrete. When the compressive loading amplitude is 12 kN, the compressive stress amplitude of 10 MPa is less than 30% of the ultimate stress of 40 MPa, so the deformation of

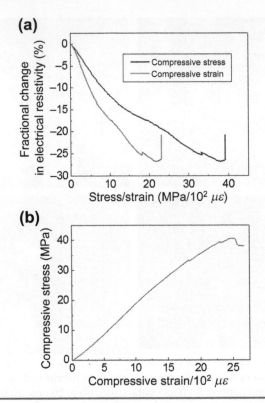

Figure 8.16 Relationships among fractional change in electrical resistivity, compressive stress, and compressive strain of concrete with hybrid carbon fiber and carbon black under compressive load [12]. (a) Fractional change in electrical resistivity at different compressive stress/strain, (b) compressive strain at different compressive stress.

concrete shown in Figure 8.16 is within an elastic regime. As shown in Figure 8.18, the compressive loading amplitude of 24 kN causes the compressive stress amplitude of 20 MPa to be more than 30% of the ultimate stress, so the deformation has gone beyond the elastic regime. As a result, the response of the contact resistance to elastic deformation is stable and reversible [21,25,26], while its response to plastic deformation is irreversible.

8.5 Performance of Carbon-Fiber-Based Self-Sensing Concrete Sensors

It can be seen from this analysis that standard sensors (as shown in Figure 8.19) can be made based on the sensing property of concrete with hybrid carbon fiber and carbon black within the elastic regime.

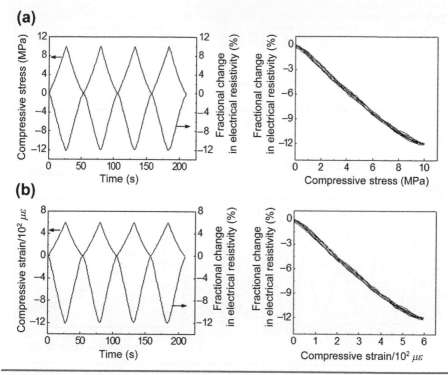

Figure 8.17 Compressive stress/strain and fractional change in electrical resistivity of concrete with hybrid carbon fiber and carbon black under repeated compressive load with amplitude of 12 kN [12].

As shown in Figure 8.17, the relationship between the fractional change in electrical resistivity and the compressive stress/strain has good linearity when the compressive stress is below 8 MPa.

The relationship between the fractional change in electrical resistivity and the compressive stress/strain shown in Figure 8.20 can be expressed as

$$\Delta\rho = -1.35\sigma \quad \text{or} \quad \Delta\rho = -0.0227\varepsilon \tag{8.11}$$

where σ is the compressive stress (MPa), and ε is the compressive strain ($\mu\varepsilon$).

And such parameters of carbon fiber concrete sensors as input, output, sensitivity, linearity, repeatability, and hysteresis are tabulated in Table 8.12. As shown in Table 8.12, the sensitivity of carbon fiber concrete sensors is 0.0227% ($\mu\varepsilon$) (a gauge factor of 227), and this means the sensitivity of sensors is well above those of conventional metal strain gauges (gauge factors of 2–3) [27].

Sensitivity S is the ratio of output increment ΔY to input increment ΔX, meaning the slope of regression curve, and is denoted by

$$S = \Delta Y/\Delta X \tag{8.12}$$

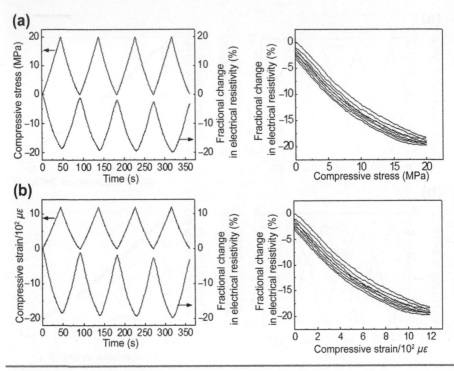

Figure 8.18 Compressive stress/strain and fractional change in electrical resistivity of concrete with hybrid carbon fiber and carbon black under repeated compressive load with amplitude of 24 kN [12].

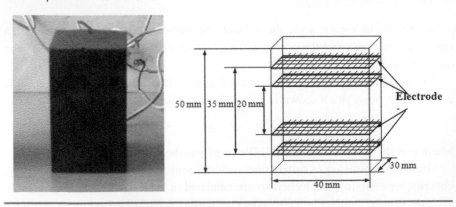

Figure 8.19 Photograph of standard sensors [12].

Linearity E indicates the offset between the input–output relationship curves and fitted regression line and is defined as

$$E = (\Delta_{max}/Y_{F.S}) \times 100\% \tag{8.13}$$

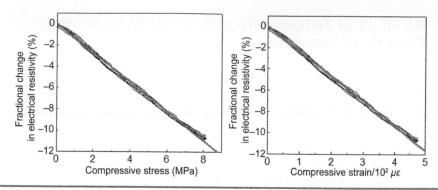

Figure 8.20 Relationships between compressive stress/strain and fractional change in electrical resistivity of carbon fiber concrete sensors [12].

where $\Delta_{\max}$ is the maximum deviation of the input–output relationship curves and the fitted regression line $Y_{F \cdot S}$ is the output range.

Repeatability R indicates the repeat degree of outputs and is denoted by

$$R = \frac{\Delta R}{Y_{F \cdot S}} \times 100\% \qquad (8.14)$$

where ΔR is the maximum repeat difference. The repeat difference is the deference of outputs corresponding to the same input in the same stroke under repeated measurement.

Hysteresis H indicates the difference of outputs corresponding to the same input during input decreasing and input increasing and is calculated by

$$H = (\Delta Y_{\max}/Y_{F \cdot S}) \times 100\% \qquad (8.15)$$

where $\Delta Y_{\max}$ is the maximum difference in the measurement range [28,29].

TABLE 8.12 Parameters of PCSS

Parameters	Values	
	Stress Sensor	Strain Sensor
Input	0 ~ 8 Mpa	0 ~ 476 $\mu\varepsilon$
Output	0 ~ 10.8%	0 ~ 10.8%
Sensitivity	1.35% Mpa	0.0227% $\mu\varepsilon$
Linearity	4.17%	4.16%
Repeatability	4.05%	4.06%
Hysteresis	3.61%	3.62%

8.6 Effect of Temperature and Humidity on Sensing Property of Sensors

Environmental factors including temperature and humidity inevitably have their effect on sensors used under realistic field conditions. To study the response of sensors to temperature, sensors were sealed to avoid the coupling influence of humidity and tested from −30 to 50 °C. The influence of humidity on the response of sensors was indirectly studied by studying the influence of water content on the response of sensors, which was measured at 20 °C and a relative humidity of 25% for different drying times from 24 h to 30 days.

8.6.1 INFLUENCE OF TEMPERATURE ON ZERO OUTPUT OF SENSORS

As shown in Figure 8.21, the output of sensors decreases as temperature increases from −30 to 30 °C and increases as temperature further increases from 30 to 50 °C. The output can go up to 6% maximum.

8.6.2 INFLUENCE OF WATER CONTENT ON ZERO OUTPUT OF SENSORS

As shown in Figure 8.22, the output of sensors decreases as drying time increases and it stabilizes in 5 days. The output can go up by 30% maximum.

For concrete with carbon fibers only, the increase in water content improves ionic conduction to reduce their electrical resistivity [30,31]. However, concrete with hybrid carbon fiber and carbon black under study in this paper exhibits an opposite trend shown in Figure 8.22, which indicates that the changes in electrical resistivity of concrete with hybrid carbon fiber and carbon black are mainly caused by carbon black, because the water absorption of carbon black can increase the contact resistance between carbon black particles and decrease the electronic conduction and the

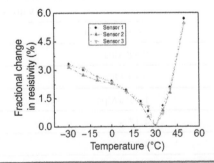

Figure 8.21 Fractional change in resistivity of sensors at different temperatures [14].

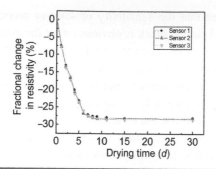

Figure 8.22 Fractional change in resistivity of sensors at different drying times [14].

positive hole conduction in carbon black particles [32]. As a result, the electrical resistivity of concrete with hybrid carbon fiber and carbon black increases as the water content in it increases. In addition, the change in the water content in the concrete with hybrid carbon fiber and carbon black in the soaking process is greater than that in the drying process, so the change in the electrical resistivity under the former condition is more obvious.

8.6.3 EFFECT OF WATER CONTENT ON SENSITIVITY OF SENSING PROPERTY OF SENSORS

Figure 8.23 shows the relationship curves between compressive strains and fractional changes in electrical resistivity of concrete with hybrid carbon fiber and carbon black at different lengths of drying time. It can be seen from Figure 8.23 that the electrical resistivity of concrete with hybrid carbon fiber and carbon black decreases with compressive strain and the fractional change in electrical resistivity corresponding to maximum strain changes from 65% to 5.5% when drying time varies from 1 day to

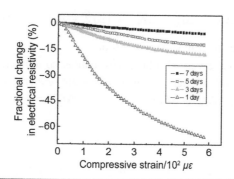

Figure 8.23 Fractional changes in electrical resistivity of concrete with hybrid carbon fiber and carbon black under compressive loads at different drying times [15].

7 days, which indicates that the sensitivity of sensing property of concrete with hybrid carbon fiber and carbon black is obvious under the effect of water content in itself. The sensing property of concrete with hybrid carbon fiber and carbon black results from the variation in the contact resistance among conductive fillers and between conductive fillers and cement matrix under compressive load. In addition, the water absorbability of carbon black can reduce the contact resistance between carbon black particles. However, the contacting and tunneling conduction effect can be enhanced under compressive loading, which improves the electronic conduction and the positive hole conduction in concrete with hybrid carbon fiber and carbon black. As a result, the sensitivity of sensing property of concrete with hybrid carbon fiber and carbon black is enhanced as the water content increases.

8.6.4 TEMPERATURE AND HUMIDITY COMPENSATION OF SENSORS

We can see that the temperature and humidity both have significant influence on the output of sensors. An effective way to eliminate the influence of temperature and humidity on the output of sensors is to apply a compensation circuit [28,29], in which the output of measurement sensor A can be given by

$$f_A = \Delta f_S + \Delta f_A^T + \Delta f_A^H + \Delta f_A^O \tag{8.16}$$

where Δf_S, Δf_A^T, Δf_A^H and Δf_A^O are the changes in outputs caused by strain, temperature, humidity, and other factors, respectively.

The output of compensation sensor B can be expressed as

$$f_B = \Delta f_B^T + \Delta f_B^H + \Delta f_B^O \tag{8.17}$$

where Δf_B^T, Δf_B^H and Δf_B^O are the changes in outputs caused by temperature, humidity, and other factors respectively.

From Eqs (8.16) and (8.17)

$$f_C = f_A - f_B = \Delta f_S + \left(\Delta f_A^T - \Delta f_B^T\right) + \left(\Delta f_A^H - \Delta f_B^H\right) + \left(\Delta f_A^O - \Delta f_B^O\right) \tag{8.18}$$

where f_C is the output of compensation circuit.

Sensors A and B have very similar performance, so temperature, humidity, and other factors have similar influence on their outputs under similar application environments, and so, $\left(\Delta f_A^T - \Delta f_B^T\right) \approx 0$, $\left(\Delta f_A^H - \Delta f_B^H\right) \approx 0$ and $\left(\Delta f_A^O - \Delta f_B^O\right) \approx 0$. Therefore

$$f_C \approx \Delta f_S \tag{8.19}$$

The outputs f_C of sensors with temperature and humidity compensations are as tabulated in Tables 8.13 and 8.14, which show that the used compensation circuit can eliminate the influence of temperature and humidity on the output of sensors.

TABLE 8.13 Output of Sensors with Temperature Compensation

Temperature (°C)		40	30	20	10	0	−10	−20
Output values after compensation	Sensor 1	0.06%	0	0.13%	0.12%	0.04%	0.02%	−0.11%
	Sensor 2	−0.23%	0	−0.03%	0.07%	−0.09%	−0.17%	−0.40%
	Sensor 3	0.27%	0	0.16%	0.05%	0.13%	0.19%	0.29%

8.7 Self-Sensing Concrete Components Embedded with Sensors

8.7.1 MAKING AND TESTING SELF-SENSING CONCRETE COMPONENTS

The materials used to make concrete components are mainly cement (Portland cement from Harbin cement factory), sand, stone, and steel, and the mix proportion for concrete is as shown in Table 8.15.

The sectional dimension and the span length of concrete beams are 150×300 mm and 2700 mm, respectively. The beams are instrumented with three linear variable differential transformers at supports and midspan to monitor deflection. The beams are subjected to third-point flexural bending testing (pure bending), as illustrated in Figure 8.24.

The space l_1 between loading point and support and the space l_2 between loading point and loading point both are 800 mm. Strain gauges (S1, S2, and S3) are located from the top down at one side of the beam midspan and measure mean strains of concrete at axle wire of sensors, neutral axis of beams, and symmetrical position of sensors' axle wire. Smart cement paste standard stress/strain sensors are placed in the middle of the top compressive region in the pure bending section that is away from loading points, as illustrated in Figure 8.25. The load is applied to the beams step by step by means of a hydraulic jack and is measured with a load cell.

The sectional dimension and the height of concrete columns are 100×100 mm and 300 mm, respectively. Smart cement paste standard stress/strain sensors are placed in the middle of concrete columns and both axle wires are concurrent. The

TABLE 8.14 Output of Sensors with Humidity Compensation

Drying Time (d)		0	1	2	3	4	5	6
Output with compensation	Sensor 1	0	0.06%	−0.45%	−0.77%	−0.52%	0.11%	0.75%
	Sensor 2	0	0.35%	0.38%	−0.36%	0.59%	0.49%	−0.06%
	Sensor 3	0	−0.29%	−0.83%	−0.40%	0.24%	−0.39%	0.81%

TABLE 8.15 Mix Proportion for Concrete

Component	Cement	Water	Sand	Stone
Weight (kg/m³)	480	170	510	1150

strain gauges are located at the lateral axle wire of concrete columns, as illustrated in Figure 8.26.

Compressive testing under load control is performed using a servo testing machine, and the loading testing device is shown in Figure 8.27. The load is applied to the columns step by step.

The strain in concrete was measured using a static strain instrument and the signals of sensors were measured with the self-developed data acquisition system as shown in Figure 8.13.

8.7.2 SELF-SENSING CONCRETE BEAMS EMBEDDED WITH SENSORS

Standard stress/strain sensors are calibrated before they are embedded into concrete beams, and calibration results are shown in Figure 8.28.

The moment of beams is calculated by Eq. (8.20).

$$M = \frac{P}{2}l_1 \tag{8.20}$$

where M is moment (N m), P is load (N), and l_1 is the space between loading point and support.

Without considering size effect of sensors, a sensor is regarded as a point in the section of beams so that the compressive stress at axle wire of sensors can be used to

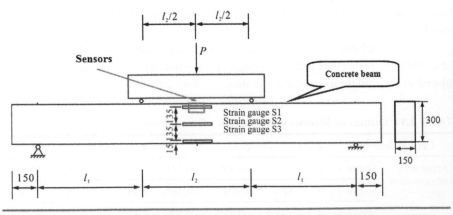

Figure 8.24 Testing sketch map for beams [14].

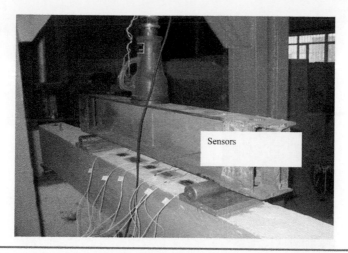

Figure 8.25 Loading photographs of beams [14].

substitute the average compressive stress of the whole section of sensors. The concrete at axle wire of sensors in the pure bending region is nonexistent and is replaced by sensors, so when the sensors and concrete are homogeneous, it bears the same compressive force as that borne by sensors during loading, and their compressive stress can be calculated by Eq. (8.21).

$$\sigma = \frac{M}{bh^3/12} \times \frac{h - h_c}{2} \tag{8.21}$$

where b is beam breadth (mm), h is beam depth (mm), and h_c is sensor depth (mm).

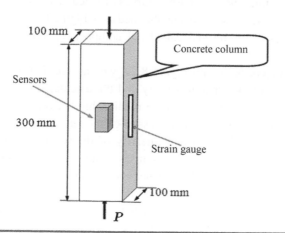

Figure 8.26 Testing sketch map for columns [14].

Column

Figure 8.27   Loading photograph of columns [14].

From Eqs (8.20) and (8.21), we can get Eq. (8.22). And the compressive stress at axle wire of sensors can be calculated by Eq. (8.22).

$$\sigma = \frac{12Pl_1(h - h_c)}{bh^3} \tag{8.22}$$

According to the compressive stress calculated by Eq. (8.22), the compressive strain of concrete at axle wire of sensors measured by strain gauge and the sensors' change rate of electrical resistivity measured by the self-developed data acquisition system during loading, the relationship curves of calculated compressive stress and change rate of electrical resistivity are obtained as in Figure 8.29 and the relationship curve of compressive strain and calculated compressive stress is obtained as in Figure 8.30.

Comparing Figure 8.28 with Figure 8.29, we can see that the relationship between change rate of electrical resistivity and the calculated compressive stress at axle wire of sensors is different from the relationship between change rate of electrical resistivity and the compressive stress of calibrated sensors. The former curves change more slowly in forepart and linearity of the whole curve decreases. The sensors' change rate of electrical resistivity is about 6% when calculated compressive stress is up to 10 MPa, and it reaches the maximum of sensors' output when calculated compressive stress is up to15 MPa. Because of the difference between calibration curves and calculated curves, the local compressive stress of concrete beams cannot be monitored by the measured electrical signals of smart cement paste standard stress/strain sensors.

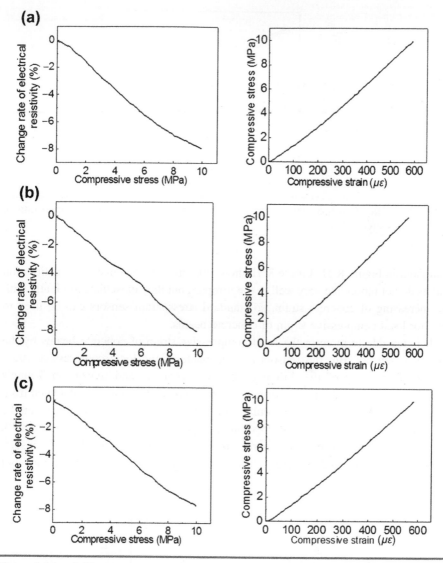

Figure 8.28 Calibration curves of standard stress/strain sensors [11]. (a) Sensor 1, (b) Sensor 2, (c) Sensor 3.

The relationship between compressive strain and change rate of electrical resistivity of calibrated sensors is obtained from the relationship between compressive strain and compressive stress and the relationship between compressive stress and change rate of electrical resistivity of calibrated sensors. And from Figure 8.28 and Figure 8.30, the relationship between compressive strain of concrete and change rate of electrical resistivity of sensors can be obtained. The obtained two relationships are

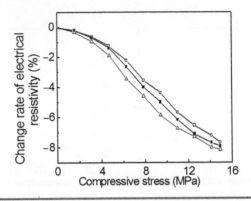

Figure 8.29 Relationship between calculated compressive stress and change rate of electrical resistivity [11].

compared in Figure 8.31. Figure 8.31 shows that monitoring curves and calibration curves do not inosculate very well in the forepart, but they inosculate gradually with the increasing of concrete strain, so standard stress/strain sensors can be used to monitor local compressive strain of concrete beams.

It is feasible to monitor the local compressive strain of concrete beams by the presented sensors. But the sensors are incapable of the local compressive stress monitoring because of the difference between concrete and sensors in Young's modular. Young's modular of sensors is about 17 GPa, which can be calculated from Figure 8.28. While Young's modular of concrete is about 25 GPa that can be calculated from the compressive stress and compressive strain curves in Figure 8.30. Because Young's modular of concrete is one-third higher than that of sensors, embedded sensors and concrete beams are coordinated in deformation, but

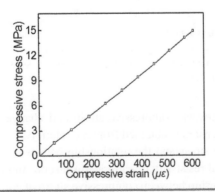

Figure 8.30 Calibration curves of standard stress/strain sensors [11].

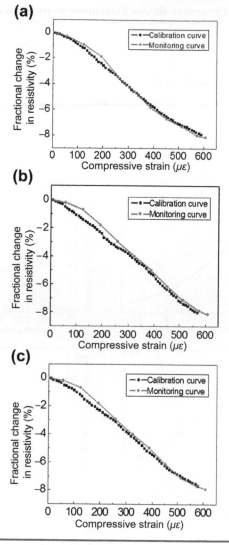

Figure 8.31   Monitoring and calibration curves established with sensors embedded in beams [11]. (a) Sensor 1, (b) Sensor 2, (c) Sensor 3.

nonsynchronous in bearing force during loading. Accordingly, standard stress/strain sensors are only competent with monitoring the local compressive strain of compressive region of concrete beams. The reason why the relationship curves of compressive strain and change rate of electrical resistivity do not inosculate very well in the forepart is that the deformation coordination of sensors and concrete needs an adjustment process.

8.7.3 Self-Sensing Concrete Beams Embedded with Sensors

Standard stress/strain sensors are calibrated before embedded into concrete columns and calibration results are shown in Figure 8.32.

The compressive stress of concrete columns is calculated by Eq. (8.23).

$$\sigma = \frac{P}{S} \tag{8.23}$$

where P is load (N) and S is sectional area.

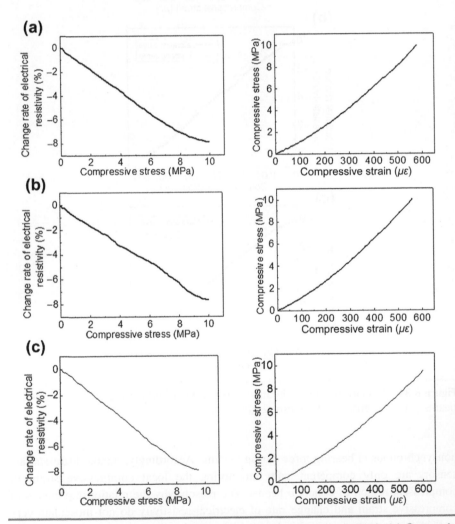

Figure 8.32 Calibration curves of standard stress/strain sensors [11]. (a) Sensor 1, (b) Sensor 2, (c) Sensor 3.

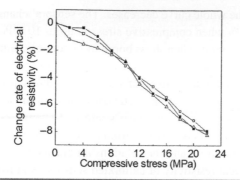

Figure 8.33 Relationship between compressive stress of columns and change rate of electrical resistivity [11].

According to the compressive stress of concrete columns, the compressive strain of concrete columns measured by strain gauge, and the sensors' change rate of electrical resistivity measured by the self-developed data acquisition system during loading, the relationship curves of compressive stress of columns and change rate of electrical resistivity and the relationship curve of compressive strain and compressive stress of columns both are obtained and are shown in Figure 8.33 and Figure 8.34 respectively.

Figure 8.32 and Figure 8.33 show that the relationship between change rate of electrical resistivity and the compressive stress of concrete columns is different from the relationship between change rate of electrical resistivity and the compressive stress of calibrated sensors. The former curves change more slowly in the forepart

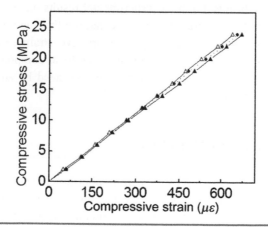

Figure 8.34 Relationship between compressive strain and compressive stress of columns [11].

and the linearity of the whole curve decreases. The sensors' change rate of electrical resistivity is about 3% when compressive stress is up to 10 MPa and it reaches the maximum of sensors' output when stress borne by concrete columns is up to 22 MPa. Since calibration curves and measured curves are different, it is impractical to monitor the local compressive stress of concrete columns just by the measured electrical signals of smart cement paste standard stress/strain sensors.

The relationship between compressive strain and change rate of electrical resistivity of calibrated sensors is obtained from the relationship between compressive strain and compressive stress and the relationship between compressive stress and change rate of electrical resistivity of calibrated sensors. And from Figure 8.32 and Figure 8.34, the relationship between compressive strain of concrete and change rate of electrical resistivity of sensors can be obtained. The obtained two relationships are compared in Figure 8.35. Figure 8.35 shows that measured curves and standardized curves inosculate gradually with the increasing of concrete strain, though they do not inosculate very well in the forepart, so it feasible for smart cement paste standard stress/strain sensors to monitor the local compressive strain of concrete columns.

From Figure 8.32 and Figure 8.34, we can calculate Young's modular of smart cement paste is about 17 GPa while that of concrete is about 37 GPa. The latter is one-half higher than the former, which decides embedded sensors and concrete columns are coordinated in deformation, but nonsynchronous in bearing force during loading. The difference of Young's modular between concrete and sensors is the reason why concrete columns with the presented sensors can realize compressive strain self-monitoring but can't realize stress self-monitoring. However, in Figure 8.33, the relationship curves of compressive strain and change rate of electrical resistivity do not inosculate very well at first, which is also because the deformation coordination of sensors and concrete needs a process to adjust just as in self-sensory concrete beams. From the experimental results on concrete beams and columns with standard stress/strain sensors, we can see that the presented standard stress/strain sensors can realize the local compressive strain self-monitoring of concrete beams and columns. In order to realize the local compressive stress self-monitoring of structures, Young's modular of sensors need be adjusted to match Young's modular of structural concrete.

8.8 Summary and Conclusions

This chapter systematically discusses carbon-fiber-based self-sensing concrete, including its fabrication, sensing properties, measurement methods, the effects of environmental parameters, and application in structural health monitoring.

Two electrode fixing styles for measuring the resistance of carbon fiber concrete are compared and the conception of embedded gauze electrode is put forward. The

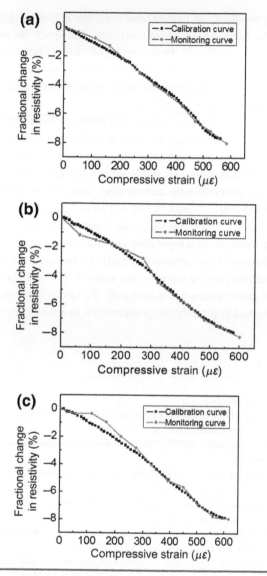

Figure 8.35 Monitoring and calibration curves established with sensors embedded in columns [11]. (a) Sensor 1, (b) Sensor 2, (c) Sensor 3.

characteristics, layout styles, and configuration parameters of embedded gauze electrode and the methods for measuring the resistance of carbon fiber concrete are studied. The data acquisition system based on voltage signal is developed. It is revealed that the four-pole layout is suitable for setting embedded gauze electrode and the direct-current four-pole method adapts to measure the resistance of carbon fiber concrete. The resistivity of carbon fiber concrete is independent of the area of voltage pole and the mesh size of gauze electrode. The space between current pole and voltage pole does not influence the resistivity of carbon fiber concrete when more than a critical value. The developed data acquisition system has many advantages such as simple circuit, good practicability, high precision, and ability to realize real-time, on-line, and multi-channel acquisition.

Carbon black is used to improve the reproducibility of sensing property of carbon fiber concrete. The sensing property of concrete with both carbon fiber and carbon black under single compressive loading and repeated compressive loads at different loading amplitudes is reversible and stable within the elastic regime. The sensors made of self-sensing concrete with carbon fiber and carbon black can achieve a sensitivity of 1.35% MPa/227, linearity of 4.17%/4.16%, repeatability of 4.05%/4.06%, and hysteresis of 3.61%/3.62% in the range 0 MPa/$\mu\varepsilon$ to 8 MPa/476$\mu\varepsilon$ of compressive stress/strain. Both temperature and humidity have significant influence on the output of sensors. A compensation circuit can be used to reduce the effect of temperature and humidity on the output of the sensors. The self-sensing experiments on concrete beams and columns embedded with the stress/strain sensors reveal that the sensors can be used to monitor the compressive strain of concrete components.

References

[1] http://www.chem.wisc.edu.
[2] Lee HG, Kim SS, Lee DG. Effect of compacted wear debris on the tribological behavior of carbon/epoxy composites. Compos Struct 2006;74(2):136–44.
[3] http://ceramics.org/ceramictechtoday/2012/09/26/the-discovery-of-carbon-fibers-and-the-practical-side-of-phase-diagrams/.
[4] Vossoughi F. Electrical resistivity of carbon fiber reinforced concrete. Berkeley, USA: Department of Civil Engineering, University of California; 94720.
[5] http://www.engr.utk.edu/mse/Textiles/CARBON%20FIBERS.htm.
[6] http://www.zoltek.com/applications/infrastructure/.
[7] Chen PW, Chung DDL. Carbon fiber reinforced concrete as a smart material capable of non-destructive flaw detection. Smart Mater Struct 1993;2:22–30.
[8] Mao QZ, Zhao BY, Sheng DR, Li ZQ. Resistance changement of compression sensible cement speciment under different stresses. J Wuhan Univ Technol 1996;11:41–5.
[9] Azhari F. Cement-based sensors for structural health monitoring. Dissertation for the Master Degree of Applied Science. Canada: University of British Columbia; 2008.
[10] http://www.civil.ubc.ca/research/materials_group/index.html.
[11] Han BG. Properties, sensors and structures of pressure-sensitive carbon fiber cement paste. Dissertation for the Doctor Degree in Engineering. China: Harbin Institute of Technology; 2005.
[12] Han BG, Ou JP. Embedded piezoresistive cement-based stress/strain sensor. Sens Actuators: Physical 2007;138(2):294–8.
[13] Han BG, Guan XC, Ou JP. Electrode design, measuring method and data acquisition system of carbon fiber cement paste piezoresistive sensors. Sensors Actuators: a Physical 2007;135(2):360–9.

[14] Ou JP, Han BG. Piezoresistive cement-based strain sensors and self-sensing concrete components. J Intelligent Material Syst Struct 2009;20(3):329–36.

[15] Han BG, Zhang LY, Ou JP. Influence of water content on conductivity and piezoresistivity of cement-based material with both carbon fiber and carbon black. J Wuhan Univ Technol-Mater Sci Ed 2010;25(1):147–51.

[16] Chen Y. AgAlSn ohmic contact electronic ink with good solderability. Electron Components Mater 1997;16(6):37–40.

[17] Mao QZ, Zhao BY, Shen DR, Li ZQ. Influence of polarization on conductivity of carbon fiber reinforced cement. Chin J Mater Res 1997;11(2):195–8.

[18] Guan ZD, Zhang ZT, Jiao JS. Physical performance for inorganic material. Beijing: Tsinghua University Press; 1992.

[19] Wo DZ. Cyclopaedia of composite material. Beijing: Chemical industry Press; 2000.

[20] Guan XC, Ou JP, Ba HJ, Han BG. State of art of carbon fiber reinforced concrete. J Harbin Univ C.E. Archit 2002;35(6):55–9.

[21] Mao QZ, Zhao BY, Shen DR, Li ZQ. Study on the compression sensibility of cement matrix carbon fiber composite. Acta Mater Compos Sin 1996;13(4):8–11.

[22] Chen B, Wu KR, Yao W. Conductivity of carbon fiber reinforced cement-based composites. Cem Concr Compos 2004;26:291–7.

[23] Balta CFJ, Bayer RK, Ezquerra TA. Electrical conductivity of PE-carbon fiber composites mixed with carbon fiber. J Mater Sci 1988;23(4):1411–5.

[24] Pramanik PK. Pressure-sensitive electrically conductive nitrile rubber composites filled with particulate carbon black and short carbon fiber. J Mater Sci 1990;25(9):3848–53.

[25] Chung DDL. Self-monitoring structural materials. Mater Sci Eng 1998;22:57–78.

[26] Wen SH, Chung DDL. Defect dynamics of cement paste under repeated compression studied by electrical resistance measurement. Cem Concr Res 2001;31:1515–8.

[27] Hou TC, Lynch JP. Conductivity-based strain monitoring and damage characterization of fiber reinforced cementitious structural components. Proc SPIE 2005;5765:419–29.

[28] Sun BY. Sensor engineering. Dalian: Press of DA'LIAN institute of technology; 1999.

[29] Tao BQ, Wang N. Electric resistance strain sensors. Beijing: National Defence Industry Press; 1993.

[30] Guan XC, Han BG, Tang MH, Ou JP. Temperature and humidity variation of specific resistance of carbon fiber reinforced cement. In: Proceedings of SPIE -smart structures and materials 2005-sensors and smart structures technologies for civil, mechanical, and aerospace systems, San Diego; 2005.

[31] Wang YL, Zhao XH. Positive and negative pressure sensitivities of carbon fiber-reinforced cement-matrix composites and their mechanism. Acta Mater Compost Sin 2005;22(4):40–6.

[32] Zhang ZG. Functional composite materials. Beijing: Chemical Industry Press; 2004.

Nickel-Powder-Based Self-Sensing Concrete

Chapter 9

271

Self-Sensing Concrete in Smart Structures. http://dx.doi.org/10.1016/B978-0-12-800517-0.00009-5

9.1 Introduction and Synopsis

Nickel is a type of transition metal. It is hard and ductile, which makes it very desirable for combining with other metals. Owing to its favorable electrical, thermal, magnetic, and magnetostrictive properties, and good resistance to corrosion, nickel has been widely used as an effective component or filler to fabricate multifunctional and smart composites with metal or macromolecule material as matrix [1–6]. Spiky spherical nickel powder possesses sharp surface protrusions similar to the ultrasharp silicon nanotips or the tips used in field emission microscopy [7–11]. Charge injected into the composite will reside on the filler particles and give rise to high local fields at the tips of the extremely sharp surface features. These nanotips are likely to have field enhancement factors as high as 1000. This will favor field-enhanced emission, which is beneficial for enhancing the responses of electrical conductivity of composites to the applied external force [7–11]. The spiky spherical nickel powders were found to be a kind of excellent filler for developing rubber-based composites with large pressure-sensitive effect in 2005 [12]. The combination of excellent electrical conductance, good resistance to corrosion by caustic alkalis (cement concrete are strongly alkaline with pH of 12–13), perfect mechanical properties, and unique spiky surface morphology also makes the spiky spherical nickel powder an especially promising candidate of filler for self-sensing concrete [13].

This chapter will introduce authors' research results on nickel-powder-based self-sensing concrete, with attention to fabrication, measurement of sensing property, sensing mechanism, and sensing characteristic model. The performance and application in traffic detection of sensors fabricated with nickel-powder-based self-sensing concrete are also introduced in this chapter [13–19].

9.2 Fabrication of Nickel-Powder-Based Self-Sensing Concrete

Both nickel powders and cement particles, used to fabricate self-sensing concrete with nickel powder, are in micro- or nanoscale dimensions. This makes the uniform dispersion of nickel powders and cement particles in the concrete a challenge, since the attractive forces between the aggregates increase with the surface area of particles. Therefore, suitable raw materials and effective processing technology are required to help the dispersion of nickel powders in concrete matrix, thus enhancing the homogeneity of the composites.

9.2.1 RAW MATERIALS

The materials used to fabricate the self-sensing concrete with nickel powder include Portland cement, water, silica fume, high-performance water-reducing agent, and nickel powders. Figure 9.1 gives the scanning electron microscope (SEM) pictures of five types of nickel powders made by Inco Ltd., and the typical physical characteristics of these five types of nickel powders are listed in Table 9.1. As shown in Figure 9.1 and Table 9.1, Type 123, Type 287, and Type 255 nickel powders are spiky spherical particles with sharp nanotips on their surfaces, while Type CNS and Type 110 nickel powders are spherical particles with smooth surfaces. Silicon fume was used to improve the dispersion of nickel powders and the homogeneity of composites due to its gradation, adsorption, and separation effects. A high-performance water-reducing agent, possessing wetting, electrostatic repulsion and/or steric hindrance effects, was used to improve the dispersion of nickel powders and cement particles and enhance the workability and homogeneity of composites. Copper loop, stainless steel mesh, copper foil, and silver paint were used to produce electrodes.

9.2.2 FABRICATION METHOD

Considering the content level of nickel powders and the workability of the concrete with nickel powder, the mix proportions of concrete were designed through trial mixing. The fabrication process of self-sensing concrete with nickel powder is illustrated in Figure 9.2. The cement was first mixed with silicon fume using a mixer. Then, water, water-reducing agent, and nickel powders were added into this mixture in order during the mixing. Finally, the mixture was poured into the oiled molds with

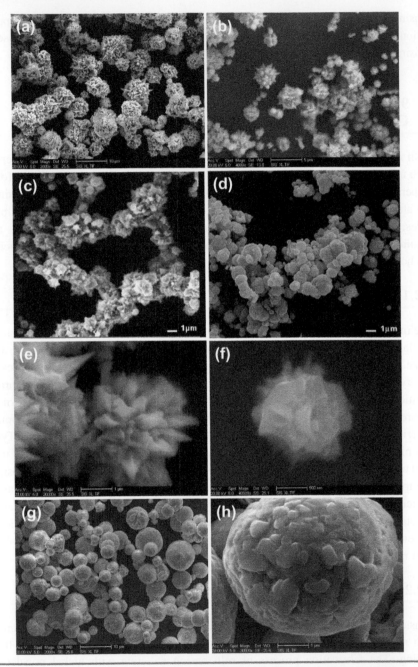

Figure 9.1 Scanning electron microscope photographs of different types of nickel powders [14]: (a) Type 123, 2000X, (b) Type 287, 4000 X, (c) Type 255, (d) Type 110, (e) Type 123, 20,000 X, (f) Type 255, 40,000X, (g) Type CNS, 2000X, (h) Type CNS, 20,000 X.

TABLE 9.1 Typical Physical Characteristics of Nickel Powders

Types of Nickel Powders	Average Size (μm)	Density (g/cm³)	Electrical Resistivity (Ω cm)	Shape
Type 123	3.0–7.0	8.9	6.84×10^{-4}	Spiky spherical
Type 287	2.6–3.3	8.9	6.84×10^{-4}	Spiky spherical
Type 255	2.2–2.8	8.9	6.84×10^{-4}	Spiky spherical
Type CNS	2.0–7.0	8.9	6.84×10^{-4}	Spherical
Type 110	0.8–1.5	8.9	6.84×10^{-4}	Spherical

two or four mesh/loop electrodes. A vibrator was then used to reduce air bubbles of the mixture. The samples, illustrated in Figure 9.3, were unmolded in 24 h and cured at a suitable temperature and relative humidity for at least 28 days. They were then dried at a temperature below 50 °C before testing. In addition, for the cases of electrical resistance measurement for the samples, two sides of some samples were pasted with copper foils using silver paint.

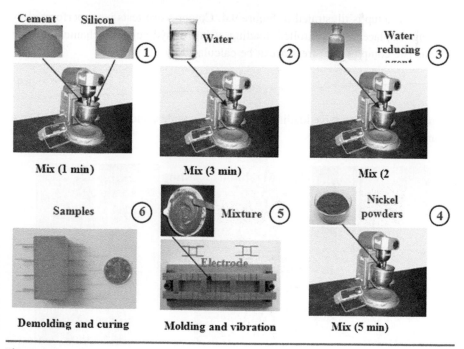

Figure 9.2 Fabrication process of samples of self-sensing concrete with nickel powder.

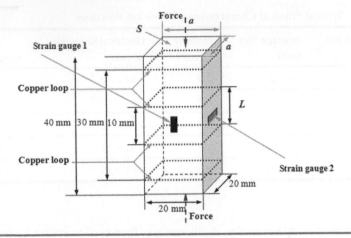

Figure 9.3 Structure of samples and arrangement of strain gauge.

9.3 Measurement of Sensing Properties of Nickel-Powder-Based Self-Sensing Concrete

9.3.1 EXPERIMENTAL LOADING AND STRAIN MEASUREMENT

The lab test setup is illustrated in Figure 9.4. Compressive tests were performed by applying displacement-controlled loading using a hydraulic mechanical testing system. The compressive stress σ can be calculated by

$$\sigma = \frac{F}{S} \tag{9.1}$$

where F is the compressive loading applying on samples, and S is the sectional area of samples.

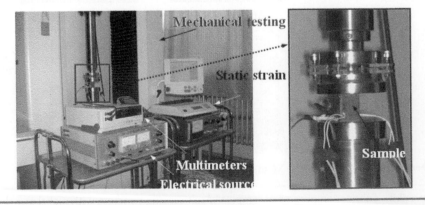

Figure 9.4 Experimental setup [13].

TABLE 9.2 Parameters of Conventional Resistive Strain Gauge

Parameters	Resistance (Ω)	Sensitivity (%)	Sensitive grid (mm × mm)
Values	120	2.12 ± 1.3	5×3

Both the longitudinal compressive strain along the stress axis of samples and the transverse tensile strain perpendicular to the stress axis of samples were measured using conventional resistive strain gauges, with their performance tabulated in Table 9.2. These strain gauges were attached right in middle on either side of the samples, as shown in Figure 9.3.

9.3.2 MEASUREMENT OF ELECTRICAL RESISTANCE

1. *Measurement methods for electrical resistance.* There are two kinds of methods for measurement of electrical resistance: the two-electrode method and the four-electrode method [20]. As shown in Figure 9.5, two copper loops in the middle were used as both current and voltage electrodes in the two-electrode method. Two copper loops in the middle and two copper loops on either side were used as voltage and current electrodes respectively in the four-electrode method 1, whereas the two copper foils were used as current electrode in the four-electrode method 2.

   As shown in Figure 9.6, the electrical resistance of the concrete with nickel powder obtained using the four-electrode method is obviously less than that obtained using the two-electrode method. The electrical resistances measured using the two types of four-electrode methods are almost the same.

   It can also be noted from Figure 9.6 that the relationship between voltage and current measured using the two-electrode method is nonlinear (i.e., the electrical resistance is not ohmic), while those measured using the four-electrode

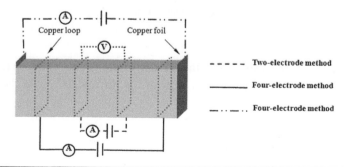

Figure 9.5 Arrangement of electrodes for measurement of electrical resistance [15].

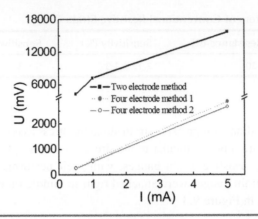

Figure 9.6 Relationship between voltage and current of samples measured with the two-electrode method or the four-electrode method [16].

method are linear (i.e., the electrical resistances are ohmic). It means that the electrical resistance measured using the four-electrode method is more accurate because the non-ohmic contact resistance between electrodes and composites/sensors in measurement using the two-electrode method is eliminated [20]. It is therefore better to use the four-electrode method to measure the electrical resistance of concrete with nickel powder. In addition, the embedded electrodes has better durability and bonding with composites than the pasted electrode, and the embedded loop electrode can greatly reduce the effect of embedded electrodes on the mechanical properties of concrete. Therefore the direct-current four-electrode method based on embedded loop electrodes was used to measure the electrical resistance of self-sensing concrete with nickel powder.

2. *Acquisition of electrical resistance signal.* The electrical resistance can be measured by using voltammetry as shown in Figure 9.7. As shown this figure, we have

$$\frac{U_{Si}}{R_{Si}} = \frac{U_C - U_{Si}}{R_{Ri}} \tag{9.2}$$

where R_{Si} and R_{Ri} are the resistance of the ith sample and the ith reference resistance, respectively. U_{Si} is the voltage at both ends of the ith sample. U_C is the external constant voltage.

Equations (9.2) can be further derived as

$$R_{Si} = \frac{U_{Si}}{U_C - U_{Si}} R_{Ri} \tag{9.3}$$

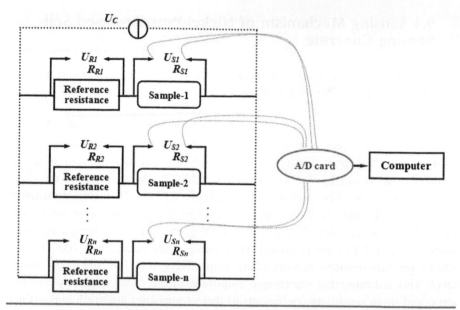

Figure 9.7 Circuit diagram of A/D acquisition and signal acquisition process [17].

If compressive loading is applied to a sample, the resistance of the sample will change. The change in resistance of the ith sample ΔR_{Si} can be expressed as

$$R_{Si} + \Delta R_{Si} = \frac{U_{Si} + \Delta U_{Si}}{U_C - (U_{Si} + \Delta U_{Si})} R_{Ri} \tag{9.4}$$

where ΔU_{Si} is the change of the voltage at both ends of the ith sample.

Since ΔU_{Si} is much smaller than $(U_C - U_{Si})$, Eqs (9.3) and (9.4) can be combined and derived as

$$\frac{R_{Si} + \Delta R_{Si}}{R_{Si}} = \frac{U_{Si} + \Delta U_{Si}}{U_C - (U_{Si} + \Delta U_{Si})} \times \frac{U_C - U_{Si}}{U_{Si}} \approx \frac{U_{Si} + \Delta U_{Si}}{U_{Si}} \tag{9.5}$$

Equation (9.5) can be further rewritten as

$$\frac{\Delta R_{Si}}{R_{Si}} \approx \frac{\Delta U_{Si}}{U_{Si}} \tag{9.6}$$

It can be seen from Eq. (9.6) that the change in electrical resistance signal of the self-sensing concrete with nickel powder caused by compressive loading is approximately equal to the change in electrical voltage signal. Therefore, the electrical resistance signal can be obtained through synchronously collecting voltage signals at both ends of the self-sensing concrete with nickel powder and reference resistance.

9.4 Sensing Mechanism of Nickel-Powder-Based Self-Sensing Concrete

9.4.1 COMPARISON OF SENSING PROPERTY OF THE SELF-SENSING CONCRETE WITH TYPE 123 AND TYPE CNS NICKEL POWDERS

Figure 9.8 shows the sensing responses of self-sensing concrete with 24 vol.% Type 123 nickel powder (with spiky surface) under uniaxial compression. It can be seen from Figure 9.8 that the relative electrical resistivity ρ/ρ_0 of these composites decreases by 0.69 maximum. In addition, the decrease of the electrical resistivity reaches 0.63 within the elastic regime. The stress sensitivity coefficient and gauge factor decrease with increase of compressive stress and strain. They can reach 0.124/MPa and 1930 (extraordinarily high compared to the values of around two for conventional resistive strain gauges and 100–200 for silicon pressure-sensitive sensors [21]), respectively, at a low stress and strain level. This indicates that the change amplitude in electrical resistivity and the stress and strain sensitivity coefficients of these composites are much higher than all of the reported self-sensing concrete mentioned above [22–24]. An interesting phenomenon observed in our experiments is the dramatic differences of electrical properties and sensitivities for self-sensing concrete containing nickel powders with different surface morphologies. Figure 9.9 gives SEM photographs of the concrete with Type 123 and Type CNS nickel powders. As shown in Figure 9.1, Table 9.1 and Figure 9.9, Type 123 and Type CNS nickel powders have similar size and similar distribution in concrete-matrix, but the surface of Type 123 nickel powder is spiky, in contrast to the smooth surface of Type CNS nickel

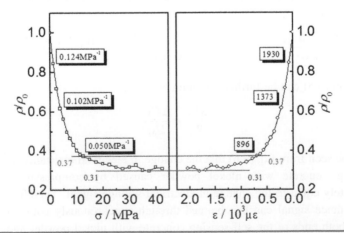

Figure 9.8 Sensing property of self-sensing concrete with Type 123 nickel powder [14].

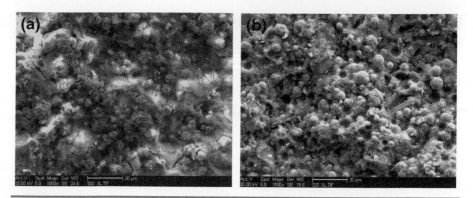

Figure 9.9 Scanning electron microscope photographs of self-sensing concrete with different nickel powders [14], (a) with Type 123 nickel powder, (b) with Type CNS nickel powder.

powder. Interestingly, the initial electrical resistivity of the concrete with Type 123 nickel powder is $2.29 \times 10^3 \, \Omega$ cm, three orders of magnitude lower than that of the concrete with the same content level of Type CNS nickel powders (about $2.36 \times 10^6 \, \Omega$ cm). The two types of concrete also show huge difference in sensing response. Unlike the strong sensing response of the concrete with Type 123 nickel powder, the sensing property of the concrete with Type CNS nickel powder cannot be obtained because the initial electrical resistivity of these composites is too large and unstable to be accurately measured under the influence of the ionic conduction in concrete-matrix [25,26]. Therefore, the differences of electrical and sensing properties between the concrete with Type 123 and Type CNS nickel powders result from the different surface morphologies of nickel particles.

9.4.2 COMPARISON OF SENSING PROPERTY OF THE SELF-SENSING CONCRETE TYPE 255 AND TYPE 110 NICKEL POWDERS

Figure 9.10 depicts the sensing responses of the self-sensing concrete with 16 vol.% Type 255 (with spiky surface) and Type 110 (with smooth surface) nickel powders under repeated compressive loading with amplitude of 2 MPa (a low stress level within the elastic regime). As shown in Figure 9.10(a), the initial electrical resistivity of the concrete with Type 255 nickel powder is $1.92 \times 10^3 \, \Omega$ cm. The electrical resistivity of the self-sensing concrete decreases reversibly during compressive loading in each cycle and increases reversibly during unloading in each cycle. The change in amplitude of the relative electrical resistivity is 0.38, i.e., the stress sensitivity coefficient of the composites is 0.19/MPa. It can be seen from Figure 9.10(b) that the initial electrical resistivity of the self-sensing concrete with Type 110 nickel powder

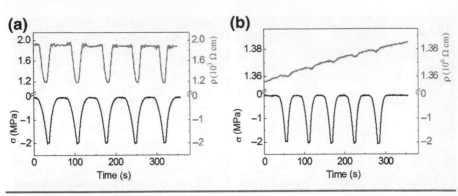

Figure 9.10 Comparison of sensing sensitivities of self-sensing concrete with Type 255 and Type 110 nickel powders under repeated compressive loading [14], (a) with Type 255, (b) with Type 110.

increases with measurement time under the influence of the ionic conduction in concrete matrix. It is about three orders of magnitude higher than that of concrete with Type 255 nickel powder. The electrical resistivity of the concrete slightly decreases upon loading and slightly increases upon unloading in every cycle under repeated loading, its original value increases in each loading and unloading cycle. The change in amplitude of the relative electrical resistivity is lower than 0.01. When the same content level of fillers is incorporated into self-sensing concrete, more conductive paths are formed in the self-sensing concrete containing fillers with smaller particle size, resulting in their lower electrical resistivity [27,28]. However, the above results are not in agreement with this general theory. Although Type 255 nickel powders have larger particle size compared with Type 110 nickel powders, as shown in Figure 9.1 and Table 9.1, the self-sensing concrete with Type 255 nickel powder presents lower and more stable electrical resistivity and more stable and sensitive response than those with Type 110 nickel powder. Therefore, the differences of electrical properties and sensing sensitivities between the self-sensing concrete with Type 255 and Type 110 nickel powders also result from the different surface morphologies of nickel particles.

9.4.3 PHYSICAL MECHANISMS OF SENSING PROPERTY OF CONCRETE WITH SPIKY SPHERICAL NICKEL POWDER

The dramatic sensing sensitivity differences of the concrete containing nickel powders with different surface morphologies lead to a questioning in the traditional conduction mechanisms of self-sensing concrete. The electrically conductive characteristics of the four types of concrete with nickel powders indicate that the electronic conduction is the dominant factor in the electrical conductivity for the concrete with Type 123 or Type 255 nickel powders, but not the dominant factor for the concrete with Type CNS or Type 110 nickel powders. In addition, according to the expected percolation threshold calculated with Monte Carlo statistical analysis [29], the nickel powder content of 16 or 24 vol.% is not high enough to set up a percolation path in the concrete-matrix. As shown in Figure 9.10, the nickel powders are randomly dispersed in the concrete-matrix and do not form a connected conductive network. The distance between two adjacent nickel powder particles inside the concrete with 24 vol.% nickel powder is calculated as follows.

When a spherical nickel powder particle with a diameter of R_n forms a sphere with a diameter of D with the concrete-matrix as shown in Figure 9.11 and these spheres distributed in the closest packing layout in concrete with nickel powder as [30]

$$\frac{V_D \times n_n}{V} = 74.05\% \tag{9.7}$$

where V_D is the volume of a sphere with a diameter of D, which can be expressed as

$$V_D = \frac{4}{3}\pi\left(\frac{D}{2}\right)^3 \tag{9.8}$$

and n_n is the number of nickel powder particles in the concrete with nickel powder with volume V, which can be expressed as

$$n_n = \frac{V_n}{V_R} = V_n\left/\left[\frac{4}{3}\pi\left(\frac{R_n}{2}\right)^3\right]\right. \tag{9.9}$$

where V_R is the volume of a nickel powder particle with a diameter of R_n, V_n is the total volume of nickel particles in the composites with volume V.

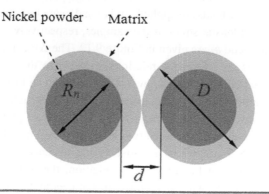

Figure 9.11   Distance d between nickel powder particles [16].

Equations (9.7), (9.8), and (9.9) can be combined into

$$\frac{D}{R_n} = \sqrt[3]{\frac{V \times 74.05\%}{V_n}} \qquad (9.10)$$

And as shown in Figure 9.11, the distance between two adjacent nickel powder particles d can be expressed as

$$d = D - R_n \qquad (9.11)$$

Removing term D from Eqs (9.10) and (9.11) yields

$$d = \left(\sqrt[3]{\frac{V \times 74.05\%}{V_n}} - 1 \right) \times R_n \qquad (9.12)$$

If the volume of nickel powder particles remains unchanged under compression, i.e., V_n is constant, change in distance between nickel particles Δd resulting from the change in volume can be given by

$$\Delta d = d - d_0 = \left(\sqrt[3]{\frac{V \times 74.05\%}{V_n}} - \sqrt[3]{\frac{V_0 \times 74.05\%}{V_n}} \right) \times R_n \qquad (9.13)$$

where V_0 and V are the volumes of samples before and after bearing compressive loads.

Fractional change in volume of samples ΔV under compression can be expressed as

$$\begin{aligned} \Delta V &= \frac{V - V_0}{V_0} = \frac{La^2 - L_0 a_0^2}{L_0 a_0^2} \\ &= \frac{(L_0 + \varepsilon_L L_0)(a_0 + \varepsilon_T a_0)^2 - L_0 a_0^2}{L_0 a_0^2} = (1 + \varepsilon_L)(1 + \varepsilon_T)^2 - 1 \end{aligned} \qquad (9.14)$$

where L_0 and a_0 are the distance between two voltage poles and the length of square side before compressive loads are applied, and ε_L and ε_T are longitudinal compressive strain and transversal tensile strain of the samples, respectively.

The measured ε_L and ε_T are given in Figure 9.12. The ΔV can be calculated from Eq. (9.14). It can be seen from the relationship between the fractional change in volume and distance between nickel powder particles shown in Figure 9.13 that the calculated distance between two adjacent nickel powder particles is about 1.30 and 3.05 µm, respectively, when the diameter of a nickel powder particle is assumed to be 3 and 7µm, so the distance between two adjacent nickel powder particles ranges from 1.3084 to 3.0529 µm, which is close to the distance between most nickel powder particles shown in Figure 9.14. In addition, the distance between nickel powder particles decreases with the volume of samples under compression. And it can be seen from Figure 9.15 that the distance between nickel powder particles

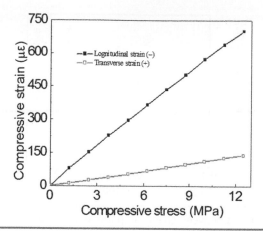

Figure 9.12 Relationship between compressive stress and strain of concrete with nickel powder [16].

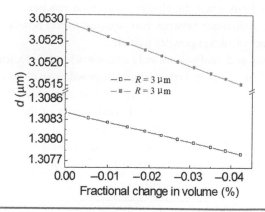

Figure 9.13 Relationship between fractional change in volume of concrete and distance between nickel powder particles [16].

Figure 9.14 Scanning electron microscope photo of concrete with 24 vol.% Type 123 nickel powder [16].

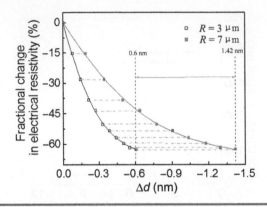

Figure 9.15 Relationship between change in particle distance and fractional change in electrical resistivity of concrete [16].

decreases by 0.6–1.4 nm when the change in volume reaches −0.042%. The 0.6–1.4 nm of decrease of distance between two adjacent nickel powder particles cannot lead to direct contact of nickel powder particles.

The above results and analysis not only explain why the electrical resistivity of the concrete with Type CNS or Type 110 nickel powders is close to that of a plain cement-matrix in a range from 10^6 to 10^9 Ω cm, but also prove that the contacting conduction does not dominate the electrical conductivity of the concrete with Type 123, Type CNS, Type 255 or Type 110 nickel powders.

As shown in Figure 9.16, the concrete with Type 123 nickel powder displays non-linear current density-electric field characteristics. Such non-linear behavior has not

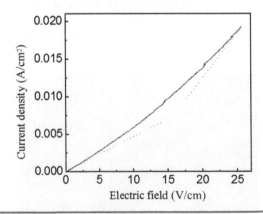

Figure 9.16 Current density-electric field characteristics of concrete with Type 123 nickel powders [14].

previously been observed in conductive concrete. Generally, the conductive concrete dominated by contacting or tunneling conduction mechanism shows ohmic behavior [31] under low measurement voltage [20,32]. It can be seen from Figure 9.17 that the nanotips on the surfaces of Type 123 or Type 255 nickel powders will result in a field-emission effect [10,11,33]. The projections on the surface of the nickel particles have tip radii below 10 nm. The local field at these sharp nanotips can have a field enhancement factor as high as 1000, which is much larger than that at the smooth surface of a spherical particle [8,12]. The high local electrical field around the sharp nanotip can decrease the barrier height and breadth on the surfaces of nickel powders, so the electrons in the nickel particles can tunnel through the surface barrier to generate field-emission effect conduction. In addition, the local high electrical field in concrete increases the potential energy of electrons through tunneling barrier between the nickel particles, which causes the enhancement in tunneling effect conduction [8,10–12,33]. Thus, the field-emission and tunneling effects lead to higher electrical conductivity of the concrete with Type 123 or Type 255 nickel powders than that of the concrete with Type CNS or Type 110 nickel powders. Furthermore, the average particle distance between nickel particles will be reduced when the concrete is deformed under compressive loading. As a result, the tunneling barrier to be transited by electrons will decrease and the field-induced tunneling can more easily occur in the composites [12,23,34,35]. Therefore, the concrete with spiky spherical nickel powders presents ultrahigh sensing responses to the applied external force.

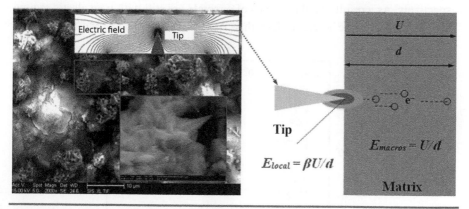

Figure 9.17 Schematic diagram of field emission in concrete with Type 123 nickel powder [14]. Note: d is the inter-particle distance, U is the applied voltage, β is the field-enhancement factor, E_{local} is the local electrical field at sharp nanotips on the surface of spiky spherical nickel powders, E_{macros} is the macroscopic electrical field.

9.5 Effect of Nickel Powder Content Level and Particle Size on Sensing Property of Concrete with Nickel Powder

9.5.1 EFFECT OF NICKEL POWDER CONTENT LEVEL ON SENSING PROPERTY

As shown in Figure 9.18, the initial electrical resistivity ρ_0 of the concrete with 22 vol.% Type 123 nickel powder is three orders lower than that of the concrete with 20 vol.% Type 123 nickel powder. While the initial electrical resistivity of the concrete with 24 vol.% Type 123 nickel powder is only about half of that of the concrete with 22 vol.% Type 123 nickel powder.

Figure 9.19 shows sensing responses of the concrete with 20, 22, and 24 vol.% Type 123 nickel powder. As can be seen in Figure 9.19, the fractional change in electrical resistivity $\Delta\rho/\rho_0$ (i.e., $(\rho - \rho_0)/\rho_0 \times 100\%$) of the concrete with 20 vol.% Type 123 nickel powder goes down to -3.60% when the compressive stress σ is less than 12.50 MPa and then begins an irregular variation with σ. $\Delta\rho/\rho_0$ of the concrete with 22 vol.% Type 123 nickel powder rapidly decreases as σ increases during the initial stage of loading and goes down to -74.67% when σ is 12.50 MPa. And then it slowly goes down to -79.82% as σ further increases. The change trend of $\Delta\rho/\rho_0$ of the concrete with 24 vol.% Type 123 nickel powder is the same as that of the concrete with 22 vol.% Type 123 nickel powder, but the change amplitude of $\Delta\rho/\rho_0$ under the same σ is lower than that of the concrete with 22 vol.% Type 123 nickel powder. According to experimental results, the ultimate compressive strengths of the concrete with 20, 22, and 24 vol.% Type 123 nickel powders are 40, 42.5, and 40 MPa, respectively. As a result, the three types of concrete are still within an elastic regime when the compressive stress is less than 12.5 MPa, just about one-third of the

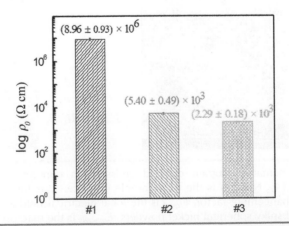

Figure 9.18   Initial electrical resistivity of concrete with different content levels of Type 123 nickel powder (#1: 20 vol.%, #2: 22 vol.%, #3: 24 vol.%) [18].

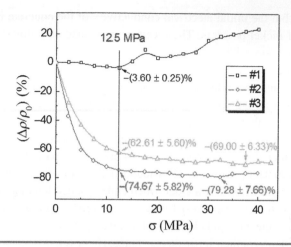

Figure 9.19 Sensing property of concrete with different content levels of Type 123 nickel powder (#1: 20 vol.%, #2: 22 vol.%, #3: 24 vol.%) [18].

ultimate stress of 40 MPa [23]. Figure 9.20 depicts the sensing sensitivity of the three types of concrete within the elastic regime. As shown in Figure 9.20, the concrete with 22 vol.% Type 123 nickel powder has the highest sensing sensitivity to compressive stress among the three types of concrete. The above results indicate that the sensing sensitivities of the concrete with 20, 22, and 24 vol.% nickel powders first increase and then decrease with the increase of nickel powder content levels.

It is interesting to note that the sensing sensitivity of the concrete with different content levels of nickel powders does not linearly increase with content levels of

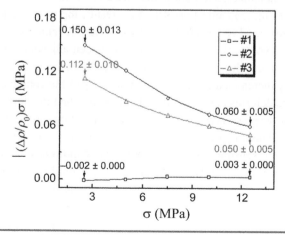

Figure 9.20 Sensing sensitivity of concrete with different content levels of Type 123 nickel powder within the elastic regime (#1: 20 vol.%, #2: 22 vol.%, #3: 24 vol.%) [18].

nickel powders, but the initial electrical conductivity of the concrete increases with content levels of nickel powders. These can be explained as follows: the electrical resistance of concrete with conductive fillers comes from two sources, i.e., the intrinsic resistance of conductive fillers and the contact resistance between conductive fillers and concrete-matrix. Thus the electrical conductivity of concrete with nickel powder strongly depends on the morphology of the nickel particle distribution and the number of contact points. The electric conductivity of individual nickel powder is 1.46×10^3 s/cm, but the contact resistance is rather complicated and depends on nickel powder geometrical feature, nickel particle distribution, and tunneling gap at contact points. The electrical resistivity of concrete with conductive fillers can be changed when they are deformed under applied loading. Several factors may contribute to the electrical resistivity change. First, when an external force is applied to the concrete with nickel powder, the nickel powder particles are deformed, resulting in the change of their intrinsic resistance. However, this resistance change is expected to be negligible because of the extremely small deformation (as the range of compressive elastic modulus of the concrete-matrix is from 7 to 28 GPa and much lower than the elastic modulus of conducting particles of 207 GPa, and the deformation of conducting particles under stress is extremely small). The second and most important factor contributing to the resistance change of the concrete is the contact resistance. Under compressive loading, the thickness of the insulating matrix between adjacent nickel particles may be changed considerably. The compressive loading gives rise to the decrease of the gap at the contact area where electrical tunneling takes place and thus decreases the contact resistance [22,32,36]. As shown in Figure 9.21(a), when the nickel powder content level is 20 vol.%, the average distance between nickel powder particles is large and the gathering of nickel powder particles is few. The conductive path is thus hard to form, even though an external force is applied to the concrete [37–39]. As a result, the concrete has high initial electrical resistivity and low sensitivity to stress. With the increase of nickel powder content level to 22 vol.%, the average distance between adjacent nickel particles in concrete decreases and the electronic transition by tunneling conduction becomes easy. This sharply increases the electrical conductivity of the concrete. Furthermore, when the concrete deforms under compressive loading, the tunneling barrier through by electrons will decrease and the field-induced tunneling can easily occur in the concrete [37–39]. These cause that the concrete presents lower initial electrical resistivity and higher sensing sensitivity. With the continuous increase of nickel powder content level to 24 vol.%, the tunneling gap would be further shortened as shown in Figure 9.21(c). And then the conductive particles become crowded and are more likely to come into contact with each other and the electrical resistivity of the concrete becomes more difficult to change under loading [37–40]. As a result, the higher content level of nickel powder induces a lower initial electrical resistivity and a lower sensitivity to stress.

Figure 9.21 Scanning electron microscope photographs of concrete with different types and different content levels of nickel powder [18], (a) with 20 vol.% of Type 123 nickel powder, (b) with 24 vol.% of Type 287 nickel powder, (c) with 24 vol.% of Type 255 nickel powder.

9.5.2 Effect of Nickel Powder Particle Size on Sensing Property

It can be seen from Figure 9.22 that the initial electrical resistivity ρ_0 of the samples decreases in turn from the composites with 24 vol.% Type 123 nickel powder, with 24 vol.% Type 287 nickel powder to those with 24 vol.% Type 255 nickel powder. The initial electrical resistivity of the concrete with 24 vol.% Type 287 nickel powder is two orders lower than that of the concrete with the same content level of Type 123 nickel powder. The initial electrical resistivity of the concrete with 24 vol.% Type 255 nickel powder is only about one-third of that of the concrete with the same content level of Type 287 nickel powder.

Figure 9.23 depicts the sensing responses and sensing sensitivity under compressive loading within the elastic regime for the concrete, respectively, with 24 vol.% Type 123 nickel powder, with 24 vol.% Type 287 nickel powder, and with 24 vol.% Type 255 nickel powder. As can be seen in Figure 9.23(a), $\Delta\rho/\rho_0$ of the three types of concrete all decreases after uniaxial compressive stress is applied. It reaches to -62.61%, -37.63%, and -9.49%, respectively, when σ is 12.5 MPa. As shown in Figure 9.23(b), the sensing sensitivities of the three types of concrete to σ are higher than 0.05/MPa, 0.03/MPa, and 0.01/MPa, respectively. It therefore can be concluded that the sensing sensitivity of the concrete with spiky spherical nickel powders decreases with particle sizes of nickel powders.

The above results can be explained as follows. When the same content level of fillers is incorporated into the concrete, more conductive channels (i.e., less the average distance between adjacent nickel particles and more stable conductive network) are formed in the concrete containing fillers with smaller particle size, resulting in their lower initial electrical resistivity and weaker response of electrical

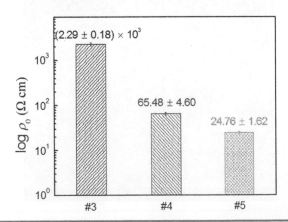

Figure 9.22 Initial electrical resistivity of concrete containing nickel powders with different particle sizes (#3: Type 123, #4: Type 287, #5:Type 255, all samples have 24 vol% of nickel powders) [18].

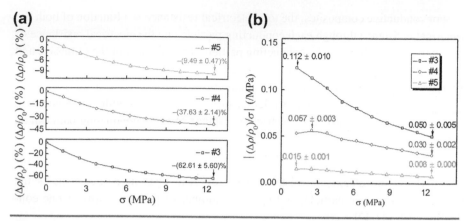

Figure 9.23   Sensing property and sensing sensitivity of self-sensing concrete containing nickel powders with different particle sizes (#3: Type 123, #4: Type 287 and #5: Type 255) [18], (a) pressure-sensitivity, (b) pressure-sensitive sensitivity.

resistivity to external force (i.e., less change in the average distance under external force and weaker response of conductive network to external force) [27,28,37–39]. As shown in Figure 9.1 and Table 9.1, the particle sizes of Type 123, Type 287, and Type 255 nickel powders decrease orderly. It can be also noted from Figure 9.1 that the smaller nickel powder particle size is, the smaller ultra-sharp tip at the surface of nickel powder particles is. The smaller ultra-sharp tip can induce a higher enhancement of the local electrical field around the tip, which will lead to stronger field-induced tunneling effect and better electrical conductivity. In addition, as shown in Figures 9.9(a), 9.21(b) and (c), the distribution density of nickel particle in the concrete with Type 287 and Type 255 nickel powders is obviously higher than that in the concrete with Type 123 nickel powder. As a result, the concrete containing nickel powders with the smallest particle size presents the lowest initial electrical resistivity and the weakest response of electrical resistivity to external force among the three types of concrete.

9.6 Sensing Characteristic Model of Nickel-Powder-Based Self-Sensing Concrete

To describe and predict the sensing characteristic behavior of the self-sensing concrete, a constitutive model to relate the change in electrical resistivity of the concrete to the applied compressive stress (or strain) should be established. The sensing characteristic model of the self-sensing concrete was set up based on the field-emission effect and the inter-particle separation change of nickel powders as follows.

For conductive composites, the total electrical resistance is a function of both the electrical resistance through each conducting particle and the electrical resistance of the matrix. Assuming that the conducting particles are spherical of the same size and dispersed randomly in matrix, the electrical resistance of the matrix is constant everywhere, and thus the number of conducting particles between the two inner electrodes becomes a factor in this relationship, as well as the number of conducting paths (see Figure 9.24(a)). The electrical resistance of every conducting path can be expressed as

$$R_i = (n_i - 1)R_m + n_iR_c \approx n_i(R_m + R_c) \tag{9.15}$$

where R_m is the electrical resistance between two adjacent particles, R_c is the resistance across one particle, and n_i is the number of particles forming one conducting path [41,42].

The total electrical resistance R of the concrete with nickel powder can then be described by

$$
\begin{aligned}
R &= \cfrac{1}{\cfrac{1}{R_1} + \cfrac{1}{R_2} + \cfrac{1}{R_3} + \cdots \cfrac{1}{R_k}} \\[2mm]
&= \cfrac{1}{\cfrac{1}{n_1(R_m + R_c)} + \cfrac{1}{n_2(R_m + R_c)} + \cfrac{1}{n_3(R_m + R_c)} + \cdots \cfrac{1}{n_k(R_m + R_c)}} \\[2mm]
&= (R_m + R_c) \times \cfrac{1}{\alpha}
\end{aligned}
\tag{9.16}
$$

where k is the number of conducting paths in the composite and $\alpha = \frac{1}{n_1} + \frac{1}{n_2} + \frac{1}{n_3} + \cdots \frac{1}{n_k}$.

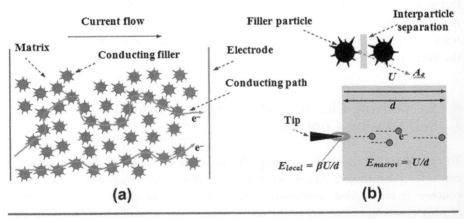

Figure 9.24 Schematic view of sensing measurement and microstructure of concrete with conductive fillers, (a) Conductive model of composite, (b) Conduction between filler particles [19].

As shown in Figures 9.1 and 9.17, the spiky, needle-like structures on the surface of nickel powder particles can result in a geometrical enhancement of the local electrical field around the tips (i.e., a field-emission effect) [12]. To model the field-emission effect aroused from the needle-like surface of nickel powder particles, Fowler-Nordheim (FN) theory is adopted in this analysis. It should be noted that the FN theory is based on the assumption of a uniform local electric field across the gap through which electrons are tunneling, which does not exactly happen on the local electrical field of nickel powder needle-like surfaces. However, since the distance between the nickel particles is much larger than the radius of the needle tips (as shown in Figures 9.1 and 9.18, the distance between nickel particles is around 5 μm while the radius of the needle tips is below 100 nm), FN theory can give reasonable good approximate calculations for this type of elongated nanostructures. Similar analyses have been performed by other researchers on the field-emission effects of similar nanostructures such as carbon nanotube tips and metallic needles [43–45]. According to the FN theory [46–48], the relationship between current density J and electrical field intensity E on the tips of nickel powder particles in the concrete (see Figure 9.24(b)) can be described as

$$J = (aE^2/\phi)\exp\left(-b\phi^{3/2}/E\right) \tag{9.17}$$

where a and b are the constants, ϕ is work function of the nickel powders, and $\exp(\cdot)$ denotes the exponential function.

Assuming that A_e is the effective cross-sectional area, where the tunneling occurs (see Figure 9.24(b)), the electrical current I can be given by

$$I = JA_e \tag{9.18}$$

As shown in Figure 9.24(b), the electrical field E can be described by [2,4–6,12]

$$E = \beta U/d \tag{9.19}$$

And then the resistance R_m is given as

$$R_m = \frac{U}{A_e J} = \frac{\phi d^2}{aA_e\beta^2 U}\exp\left(\frac{b\phi^{3/2}d}{\beta U}\right) \tag{9.20}$$

As the electrical conductivity of the conducting particle is very large compared with that of the concrete-matrix, the electrical resistance across the particle can be neglected (i.e., $R_c \approx 0$). Then substituting Eq. (9.20) into Eq. (9.16) gives

$$R = \frac{1}{\alpha} \times \left[\frac{\phi d^2}{aA_e\beta^2 U}\exp\left(\frac{b\phi^{3/2}d}{\beta U}\right)\right] \tag{9.21}$$

The electrical resistivity ρ of a sample can be given by

$$\rho = R\frac{L}{S} = \frac{1}{\alpha} \times \left[\frac{\phi d^2}{aA_e\beta^2 U}\exp\left(\frac{b\phi^{3/2}d}{\beta U}\right)\right] \times \frac{L}{S} \tag{9.22}$$

If a compressive stress is applied to the sample, the electrical resistance will be altered due to the change of inter-particle separation, which is caused by the difference in compressibility between the filler particles and matrix. Assumed that the inter-particle separation changes from d_0 to d under the applied stress, the relative electrical resistivity (ρ/ρ_0) can then be calculated by

$$\frac{\rho}{\rho_0} = \frac{d^2}{d_0^2}\exp\left[\frac{b\phi^{3/2}}{\beta U}(d - d_0)\right] \tag{9.23}$$

As the range of compressive elastic modulus of the concrete-matrix is from 7 to 28 GPa [49] and much lower than 207 GPa, the elastic modulus of conducting particles, the deformation of conducting particles under stress can be neglected. As a result, the change of inter-particle separation along the conducting path is due only to the deformation of the concrete-matrix. Therefore, if the applied stress is uniaxial press, the separation d under the applied stress can be calculated as

$$d = d_0(1 - \varepsilon) = d_0\left(1 - \frac{\sigma}{M}\right) \tag{9.24}$$

where ε is the strain of the concrete-matrix, which is equal to the strain of the composites. M is the compressive modulus of the concrete-matrix [32].

From Eqs (9.23) and (9.24), the relative electrical resistivity (ρ/ρ_0) can be represented as

$$\frac{\rho}{\rho_0} = \left(1 - \frac{\sigma}{M}\right)^2 \exp\left[-\frac{b\phi^{3/2}d_0}{\beta UM}\sigma\right]$$

$$\approx \left(1 - \frac{\sigma}{M}\right)^2\left[1 - \frac{b\phi^{3/2}d_0}{\beta UM}\sigma + \frac{1}{2}\left(\frac{b\phi^{3/2}d_0}{\beta UM}\right)^2\sigma^2 - \frac{1}{6}\left(\frac{b\phi^{3/2}d_0}{\beta UM}\right)^3\sigma^3 + \cdots\right] \tag{9.25}$$

And the fractional change in electrical resistivity $\rho(\sigma)$ can be given by

$$\rho(\sigma) = \rho(\sigma_0)\left(1 + A^*\sigma + B^*\sigma^2 + C^*\sigma^3 + D^*\sigma^4 + \cdots\right)$$
$$= \rho(\sigma_0) + A\sigma + B\sigma^2 + C\sigma^3 + D\sigma^4 + \cdots \tag{9.26}$$

where A^*, B^*, C^*, D^*, A, B, C, and D are polynomial coefficients.

TABLE 9.3 Regression Results of Relationships between ρ/ρ_0 and Compressive Stress of Concrete

Simples	Polynomial Order	Regression Coefficients				
		A	B	C	D	R^2
With Type 123 nickel powders	2	−259.37	11.93	–	–	0.9987
	3	−313.69	25.82	−0.80	–	0.9998
	4	−333.70	35.08	−2.05	0.05	0.9999
With Type 287 nickel powders	2	−3.96	0.16	–	–	0.9999
	3	−4.12	0.20	-3.00×10^{-3}	–	0.9999
	4	−4.10	0.19	-9.37×10^{-4}	-6.29×10^{-5}	0.9999
With Type 255 nickel powders	2	−0.40	0.02	–	–	1
	3	−0.44	0.03	-6.35×10^{-4}	–	1
	4	−0.43	0.02	4.47×10^{-4}	-4.45×10^{-5}	1

The experimental data on the sensing property are fitted using Eq. (9.26) with different polynomial orders. The regression coefficients are listed in Table 9.3. Figure 9.25 shows the fitted results of the composites with three kinds of nickel powders. It can be seen from Table 9.3 and Figure 9.25 that the fitting curves using the third-order and fourth-order regressions agree well with the experimental data, and use of the third-order polynomial regression is acceptable for describing the relationship between the electrical resistivity ρ and the compressive stress σ. The relative errors between bulk values calculated with the third-order polynomial model and real values, i.e., the ratio of difference between bulk values and real values to real values, are illustrated in Figure 9.26. This figure shows the absolute values of the relative errors are smaller than 3%.

In addition, from Eqs (9.23) and (9.24), Eq. (9.25) can be derived

$$\frac{\rho}{\rho_0} = (1 - \varepsilon)^2 \exp\left[-\frac{b\phi^{3/2}d_0}{\beta U}\varepsilon \right]$$

$$\approx (1 - \varepsilon)^2 \left[1 - \frac{b\phi^{3/2}d_0}{\beta U}\varepsilon + \frac{1}{2}\left(\frac{b\phi^{3/2}d_0}{\beta U}\right)^2 \varepsilon^2 - \frac{1}{6}\left(\frac{b\phi^{3/2}d_0}{\beta U}\right)^3 \varepsilon^3 + \cdots \right]$$

$$= 1 + A^{\#}\varepsilon + B^{\#}\varepsilon^2 + C^{\#}\varepsilon^3 + \cdots$$

(9.27)

where $A^{\#}$, $B^{\#}$, and $C^{\#}$ are polynomial coefficients.

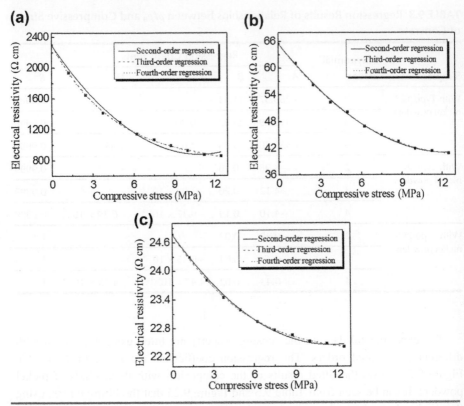

Figure 9.25 Relationships between electrical resistivity and compressive stress of concrete with nickel powder [19], (a) with Type 123 nickel powders, (b) with Type 287 nickel powders, (c) with Type 255 nickel powders.

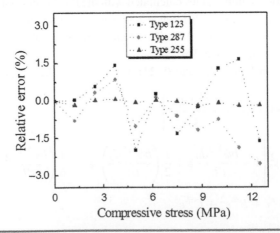

Figure 9.26 Relative errors between bulk values calculated with the third-order polynomial model and real values [19].

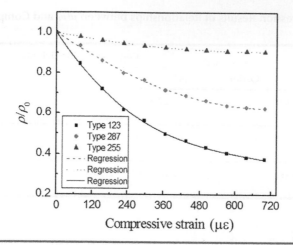

Figure 9.27 Relationships between ρ/ρ_0 and compressive strain of concrete with different nickel powders [19].

Both the fitting curves shown in Figure 9.27 and the regression coefficients listed in Table 9.4 show that a third-order polynomial regression of Eq. (9.27) can accurately fit the relationship between the relative electrical resistivity ρ/ρ_0 and the compressive strain ε of these composites.

From the above analysis, it can therefore be concluded that the third-order polynomial model can be taken as a sensing characteristic model of the concrete containing nickel powders with needle-like surface.

9.7 Nickel-Powder-Based Self-Sensing Concrete Sensors and Wireless Stress/Strain Measurement System Integrated with Them

The current price of nickel powders is so high that production of large civil structures with the concrete containing nickel powders would be cost-prohibitive. However, if the self-sensing concrete can be used in key positions of significant engineering structures (e.g., they can be developed as small-scale sensors and embedded in key positions of structures), the whole construction cost would be little increased. Moreover, the construction technology of the whole structures also would not change obviously. In addition, like many other sensors used in concrete structures, the sensing signals of the self-sensing nickel powder concrete sensors are conventionally acquired using a wired acquisition system. The connection of wires is time consuming and cost expensive. The wires may also affect the reliability of the signal transmission and are easily damaged during the placement operation or in service. Furthermore, there may be cases in which

TABLE 9.4 Regression Results of Relationships between ρ/ρ_0 and Compressive Strain of Concrete

Simples	Polynomial Order	Regression Coefficients			
		$A^{\#}$	$B^{\#}$	$C^{\#}$	R^2
With Type 123 nickel powders	3	-2.22×10^{-3}	2.85×10^{-6}	-1.39×10^{-9}	0.9999
With Type 287 nickel powders	3	-9.98×10^{-4}	5.96×10^{-7}	8.93×10^{-11}	0.9999
With Type 255 nickel powders	3	-2.86×10^{-4}	2.58×10^{-7}	-6.02×10^{-11}	1

wires cannot be placed in certain locations of a structure or even in the entire structure. The wireless measurement technique has been used for sensor signal acquisition in an attempt to solve the above-mentioned problems. A wireless measurement system integrated with the self-sensing concrete sensors as sensing element should be developed for practical applications.

9.7.1 SELF-SENSING NICKEL POWDER CONCRETE SENSORS

1. *Performance of self-sensing nickel powder concrete sensors under repeated compressive loading at an amplitude of 12.5 MPa.* The sensing response of concrete with nickel powder is stable and reversible within the elastic regime, so the concrete can serve as stress/strain sensors [50]. As shown in Figure 9.19, the composites are still within an elastic regime when the compressive stress is below 13 MPa. It can been seen from Figure 9.28 that the relationships between the compressive force and the fractional change in electrical resistivity of sensors have the same change trends when compressive loads are applied at an amplitude of 5 kN three times, which indicates that the sensing property of the sensors has good repeatability.

   The sensitivity of the sensors to compressive stress is higher than 0.050092/MPa and goes up to 0.123648/MPa, which is obviously higher than 0.0135/MPa of the sensors made of self-sensing concrete with both carbon fiber and carbon black [23] and 0.028/MPa of the self-sensing concrete with ozone-treated carbon fiber [22]. The gauge factor of the self-sensing nickel powder concrete sensors is higher than 895.450 and goes up to 1929.50, which is obviously higher than 227 of the sensors made of self-sensing concrete with both carbon fiber and carbon black [23] and 560 of the self-sensing concrete with ozone-treated carbon fiber [22]. It is nearly 400–600 times the sensitivity of a conventional resistive strain gauge. The resolution of the data acquisition system is 0.0001%, which confirms that a minimum stress resolution of 20 Pa $(0.0001\%/(0.050092/\text{MPa}) \approx 20\,\text{Pa})$ and a minimum strain

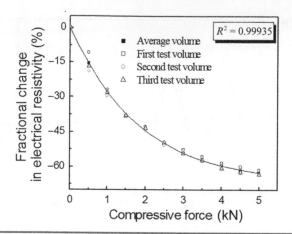

Figure 9.28 Relationship between compressive force and fractional change in electrical resistivity of sensors under three times repeated loading within elastic regime [15].

resolution of 0.001 με (0.0001%/(0.000895450/με) ≈ 0.001 με) can be achieved with these sensors. It can therefore be concluded that the self-sensing nickel powder concrete sensors have favorable sensitivity and resolution to stress/strain.

2. *Performance of self-sensing nickel powder concrete sensors under repeated compressive loading at an amplitude of 2.5 MPa.* It can be seen from Figure 9.29(a) that the fractional change in the electrical resistivity of sensors decreases under uniaxial compression. It goes up to 42.719% when the compressive stress/strain is in the range from 0 MPa/0 με to 2.5 MPa/311.5 με. As shown in Figure 9.29(a), the fractional change in electrical resistivity reversibly decreases upon loading and increases upon unloading in each cycle.

According to the linear regression, i.e., $Y = kX$, the relationship between compressive stress/strain σ/ε and the fractional change in electrical resistivity $\Delta\rho/\rho_0$ shown in Figure 9.29(b) can be expressed as

$$\Delta\rho/\rho_0 = -0.16629\sigma \quad \text{or} \quad \Delta\rho/\rho_0 = -1319.8\varepsilon \tag{9.28}$$

As shown in Table 9.5, the relationship between the compressive stress/strain and the fractional change in electrical resistivity has a good linearity. The sensitivity of the self-sensing nickel powder concrete sensors to compressive stress/strain is 16.629%/MPa/0.13198%/με, which is obviously higher than 1.35%/MPa/0.0227%/ με of the stress/strain sensors made of self-sensing concrete with carbon fiber and carbon black [23] and is nearly 650 times the sensitivity of a conventional resistive strain gauge to compressive strain.

It therefore can be concluded that the developed self-sensing nickel powder concrete sensors have a high sensitivity to compressive stress/strain.

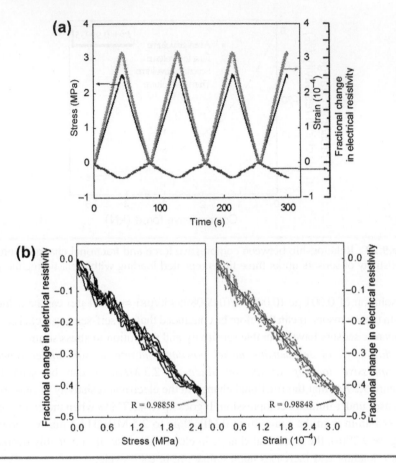

Figure 9.29 Relationship between input and output of sensors, (a) Variation of electrical resistivity, stress and strain with time, (b) Relationship between electrical resistivity and stress/strain [13].

TABLE 9.5 Regression Parameters for Relationship between Compressive Stress/Strain and Fractional Change in Electrical Resistivity of Self-Sensing Nickel Powder Concrete Sensors

Regression Parameters	Values of Parameters	
	Compressive Stress	Compressive Strain
k	16.629%/MPa	0.13198%/$\mu\varepsilon$
r (Correlation coefficient)	0.98858	0.98848

9.7.2 WIRELESS STRESS/STRAIN MEASUREMENT SYSTEM INTEGRATED WITH SELF-SENSING NICKEL POWDER CONCRETE SENSORS

1. *Wireless acquisition of output signals from self-sensing nickel powder concrete sensors.* The wired and wireless acquisition systems for output signals of the sensors are given in Figure 9.30, and the output signals of the sensors acquired using two types of acquisition systems are compared in Figure 9.31. As shown in Figure 9.31(a), the output signals of the sensors without loading acquired using both types of acquisition systems are very close to each other within the range from −0.4 to 0.4%. This indicates that the wireless acquisition system is as stable as the wired one.

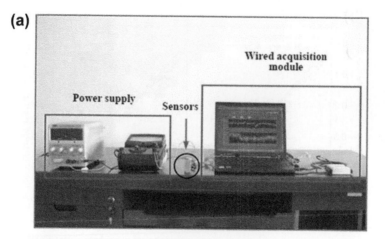

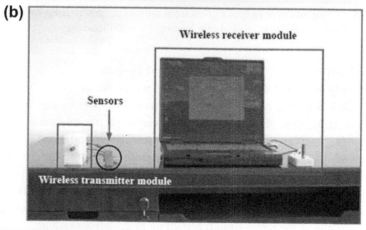

Figure 9.30 Output signal acquisition system of sensors [13], (a) wired acquisition system, (b) wireless acquisition system.

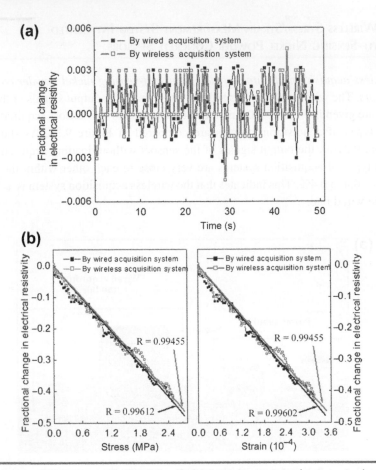

Figure 9.31 Comparison of output signals from sensors acquired using wired and wireless acquisition systems [13], (a) without loading, (b) with loading.

Figure 9.31(b) shows that the output acquired using the wireless acquisition system linearly decreases with the increase of compressive stress/strain. It decreases to 42.926% when the compressive stress/strain is 2.5 MPa/311.5 με. According to linear regression, the regression coefficients of the relationships between output (the fractional change in electrical resistivity) and input (compressive stress/strain) of the sensors are tabulated in Table 9.6. It can be seen from Figure 9.31(b) and Table 9.6 that the relationship between input and output signals of the sensors acquired using the wireless acquisition system is consistent with that acquired using the wired acquisition system when the compressive stress/strain is in the range from 0 MPa/0 με to 2.5 MPa/311.5 με.

All these results indicate that the developed wireless acquisition system is reliable for gathering output signals from the self-sensing nickel powder concrete

TABLE 9.6 Regression Parameters for Relationship between Input and Output of Sensors

	Values of Parameters			
	Acquired using Wired Acquisition System		Acquired using Wireless Acquisition System	
Regression Parameters	Compressive Stress	Compressive Strain	Compressive Stress	Compressive Strain
k	17.448%/MPa	0.13892%/με	16.894%/MPa	0.13365%/με
r (Correlation coefficient)	0.99612	0.99602	0.99455	0.99445

sensors. It is feasible to develop a wireless stress/strain measurement system integrated with these sensors.

2. *Performance of wireless stress/strain measurement system integrated with self-sensing nickel powder concrete sensors.* The relationship between input (compressive stress/strain σ/ε) and output (the fractional change in electrical resistivity $\Delta\rho/\rho_0$) of the wireless stress/strain measurement system integrated with the self-sensing nickel powder concrete sensors, as shown in Figure 9.31(b), can be expressed as

$$\Delta\rho/\rho_0 = -0.16894\sigma \text{ or } \Delta\rho/\rho_0 = -1336.5\varepsilon \qquad (9.29)$$

Table 9.7 tabulates main parameters (including input range, output range, sensitivity, and resolution) of the wireless stress/strain measurement system integrated with the self-sensing nickel powder concrete sensors. It can be seen from Table 9.7 that the sensitivity of the wireless stress/strain measurement system to compressive stress/strain is 16.894%/MPa/0.13365%/με (which means a gauge factor of 1336.5). The resolution of the wireless data acquisition system integrated

TABLE 9.7 Performance Parameters of Wireless Stress/Strain Measurement System Integrated with Self-Sensing Nickel Powder Concrete Sensors

	Values of Parameters	
Performance Parameters	Compressive Stress	Compressive Strain
Input range	0–2.5 MPa	0–311.5 με
Output range	0–42.719%	0–42.719%
Sensitivity	16.894%/MPa	0.13365%/με
Resolution	150 Pa	0.02 με

with self-sensing nickel powder concrete sensors is 0.0025%. It is confirmed that a stress resolution of 150 Pa (0.0025%/(16.894%/MPa) ≈ 150 Pa) and a strain resolution of 0.02 με (0.0025%/(0.13365%/με) ≈ 0.02 με) can be achieved using the wireless stress/strain measurement system. Therefore, the sensitivity and the resolution of the wireless stress/strain measurement system integrated with self-sensing nickel powder concrete sensors are well above those achievable with the wireless stress/strain measurement system integrated with conventional resistive strain gauges (gauge factors of 2–3 and resolution of 1 με) [51].

One of the goals to be achieved during the design of a wireless measurement system is to keep the energy consumption of the system as low as possible [51–53]. The output voltage from the self-sensing nickel powder concrete sensors is about 2.010 V in the wireless stress/strain measurement system without loading, as shown in Figure 9.32. The wireless stress/strain measurement system integrated with the self-sensing nickel powder concrete sensors has a sensitivity of 16.894%/MPa/ 0.13365%/με to compressive stress/strain. Therefore, the output signals from the sensors in the wireless stress/strain measurement system integrated with the self-sensing nickel powder concrete sensors are not necessary to be amplified for acquisition and transmission. However, the output signals from resistive strain gauges in the wireless stress/strain measurement system integrated with conventional resistive strain gauges are weak. As a result, it is necessary to have an amplification circuit for a standard output, which will result in a complex circuit and an enhancement of energy consumption [51]. In addition, the electrical resistivity of the self-sensing nickel powder concrete sensors is $5.400 \times 10^3 \, \Omega$ cm, so their electrical resistance of 1450 Ω is higher than that of conventional resistive strain gauges of 120 Ω. As a result, the power consumed by the self-sensing nickel powder concrete sensors is much lower than that consumed by conventional resistive strain gauges as the external power supply is same [51]. The analyses above indicate that the presented wireless stress/strain measurement system integrated with the self-sensing nickel powder concrete sensors features simple circuit and low energy consumption.

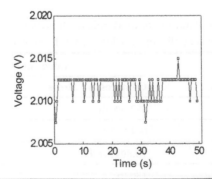

Figure 9.32 Output voltage from sensors in wireless stress/strain measurement system without loading [13].

9.8 Application of Nickel-Powder-Based Self-Sensing Concrete Sensors in Vehicle Detection

9.8.1 Sensing Property of Nickel-Powder-Based Self-Sensing Concrete Sensors Embedded in Pavement

Figure 9.33 shows the variation of the electrical resistivity ρ of the nickel-powder-based self-sensing concrete sensors under compressive loading before they are embedded in pavement. As shown in Figure 9.33, the electrical resistivity of the sensors decreases upon loading and increases upon unloading in every cycle under repeated compressive loading. It indicates that the response of electrical resistivity of the sensors to compressive stress σ is repeatable under repeated compressive loading. The fractional change in electrical resistivity $\Delta\rho/\rho_0$ reaches about 18% as the compressive stress is 0.5 MPa. According to these results, it can be concluded that the response of electrical resistivity of the sensors to compressive stress is reversible and sensitive. This means that the sensors have excellent sensing capability. Since the weight of a small vehicle is approximately 1000 kg and the area of the four tires on the ground is about 200 cm^2, the pressure on the sensor from the passing vehicle is about 0.5 MPa. Therefore, the nickel-powder-based self-sensing concrete sensors have high enough sensitivity to be a vehicle detector.

9.8.2 Detection of Passing Vehicles by Using Self-Sensing Pavement Embedded with Nickel-Powder-Based Self-Sensing Concrete Sensors

Figure 9.34 shows the road test setting. The pavement was cut and the nickel-powder-based self-sensing concrete sensors were embedded into the concrete pavement and fixed with cement paste. Figure 9.34(a) shows the sensors before and after being fixed with cement paste. The sensor array composed of eight sensors is 1.2 m in length. The spacing between two adjacent sensors is about 11 cm. The

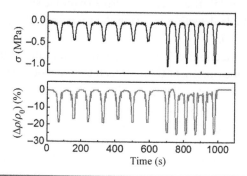

Figure 9.33 Relationship between compressive stress and fractional change in electrical resistivity of sensors [17].

Figure 9.34 Road test of self-sensing pavement [17], (a) self-sensing pavement embedded with sensor array, (b) road test.

sensor array covers one-half of the lane in width to ensure at least one sensor can be passed over by the vehicle's tire. A car was driven over the sensor array to investigate the feasibility of vehicle detection by using the proposed self-sensing pavement, as shown in Figure 9.34(b).

Vehicle passing detection tests were performed and the voltage time-histories were collected. The measurement time of each test is about 180 s. To decrease the effect of measurement noise, a low-pass filter with cut-off frequency of 25 Hz was used to preprocess the measured voltage signals. The low-pass filtered and the measured voltage signals are illustrated in Figure 9.35. As shown in this figure, abrupt changes present in the voltage signal curves when the vehicle passes over the sensor each time. A comparison between the measured data and the hand-recorded data are made to confirm that the time points of abrupt changes in the voltage

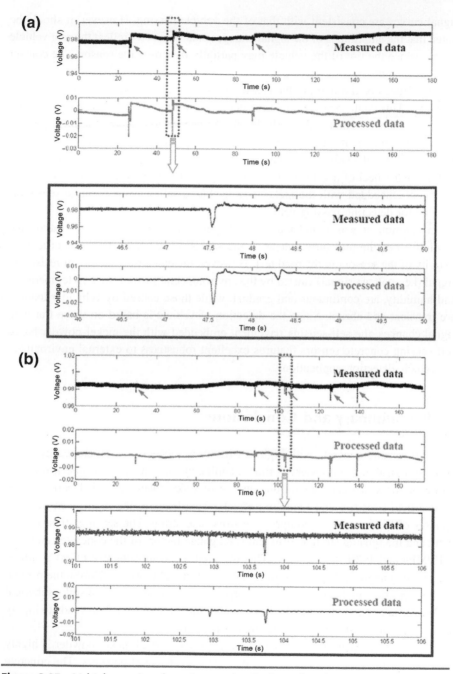

Figure 9.35 Vehicle passing detection results. Each peak indicates a passing vehicle and the zoom-out figures show the signal changes from the front wheel and the rear wheel of a vehicle [17], (a) test 1 data, (b) test 2 data.

signal curves are consistent with that of the vehicle passing. However, it should be noted that the amplitudes of the signal changes cannot exactly reflect the real vehicle weight, since the tire of the vehicle may partially pass over the sensor. The contact area between the passing vehicle tire and the sensors is not equal each detection, so the amplitudes of the abrupt changes in voltage signals are different even for the same car. This could be acceptable for detecting traffic flow rates, vehicular speed, and traffic density, since only peak signals are needed to evaluate the three parameters. Furthermore, it can be seen from the two detailed zoom-out drawings in Figure 9.35 that the self-sensing pavement can detect the passing of the front wheel and the rear wheel of a vehicle. These findings indicate that the use of self-sensing pavement embedded with the nickel-powder-based self-sensing concrete sensors is feasible for detecting passing vehicles.

In addition, it was found that the changes in environmental temperature and humidity (i.e., the moisture content of the sensors) had effect on the electrical resistivity of the sensors in the road test. However, the changes in electrical resistivity signal (i.e., voltage signal) caused by the environmental factors, such as temperature and humidity, are continuous and gradual, while those caused by vehicular loading are transient and abrupt. Since the detection of vehicle is based on those transient signal changes, the self-sensing pavement embedded with the nickel-powder-based self-sensing concrete sensors features excellent robustness to external environment for vehicle detection applications.

9.9 Summary and Conclusions

The self-sensing concrete with high sensitivity is successfully developed by using spiky spherical nickel powders as fillers. Unlike the normal smooth spherical nickel particles, the sharp nanotips on the surface of spiky spherical nickel particles can induce field-emission and tunneling effects. This leads to the highly sensitive responses of the concrete to compressive stress and strain. The sensing stress coefficient and the sensing strain coefficient (i.e., gauge factor) of the concrete can reach 0.19/MPa and 1930, respectively. This ultrahigh pressure-sensitive effect is attributed to the unique spiky surface morphology of nickel powders. Unlike the normal smooth spherical nickel particles, the sharp nanotips on the surface of spiky spherical nickel particles can induce field-emission and tunneling effects. This leads to highly sensitive responses to compressive stress and strain.

The sensing sensitivity of concrete with spiky spherical nickel powder is highly concerned with the content level and particle size of nickel powders. The concrete with 22 vol.% of nickel powder shows the highest sensing sensitivity among the concrete containing 20, 22, and 24 vol.% of nickel powder with particle sizes in the range of 3–7 μm. At 24 vol.% of nickel powder content level, the concrete containing nickel powder with particle sizes in the range of $3 \sim 7\,\mu m$ displays high sensing sensitivity. The sensing characteristic model based on the field-emission effect and the inter-particle separation change of nickel powders can accurately describe the relationship between the change in the electrical resistivity and the applied compressive stress within elastic regime under uniaxial compression.

The nickel-powder-based self-sensing concrete can be used to fabricate sensors with high sensitivity to stress/strain. The wireless measurement system integrated with these sensors has such advantages as high sensitivity to stress/strain ($16.894\%/$ MPa/$0.13365\%/\mu\varepsilon$), high stress/strain resolution ($150\,Pa/0.02\,\mu\varepsilon$), simple circuit, and low energy consumption. The sensor array is embedded into a concrete pavement to form the self-sensing pavement. Due to the high sensing sensitivity of the self-sensing nickel powder concrete sensors, the designed self-sensing pavement can accurately detect the passing vehicles. These findings indicate that the self-sensing pavement embedded with the nickel powder based self-sensing concrete sensors has a great potential for traffic detection and structural health monitoring.

References

[1] Park JM, Kim SJ, Yoon DJ, Hansen G, De Vries KL. Self-sensing and interfacial evaluation of Ni nanowire/polymer composites using electro-micromechanical technique. Compos Sci Technol 2007;67:2121–34.

[2] Garcia-Alonso A, Garcia J, Castano E, Obieta I, Gracia FJ. Strain sensitivity and temperature influence on sputtered thin films for piezoresistive sensors. Sens Actuators A Phys 1993; 37-38(2):784–9.

[3] Horia C, Maria U, Florin R, Cornelia H, Maria N. Ni-Ag thin films as strain-sensitive materials for piezoresistive sensors. Sens Actuators A Phys 1999;76(1–3):376–80.

[4] Ausanio G, Barone AC, Campana C, Iannotti V, Luponio C, Pepe GP, et al. Giant resistivity change induced by strain in a composite of conducting particles in an elastomer matrix. Sens Actuators A Phys 2006;127(1):56–62.

[5] Wisitsoraat A, Patthanasetakul V, Lomas T, Tuantranont A. Low cost thin film based piezoresistive MEMS tactile sensor. Sens Actuators A Phys 2007;139(1–2):17–22.

[6] Rajanna K, Mohan S, Nayak MM, Gunasekaran N, Muthunayagam AE. Pressure transducer with Au-Ni thin-film strain gauges. IEEE Trans Electron Devices 1993;40(3):521–4.

[7] Chen L. Experimental study of ultra-sharp silicon nano-tips. Solid State Commun 2007;143:553–7.

[8] Edgcombe CJ, Valdre U. Field emission and electron microscopy. Microsc Microanal 2000;6:380–7.

[9] Batelaan H, Uiterwaal K. Microscopy: tip-top imaging. Nature 2007;446:500–1.

[10] Meyyappan M. Carbon nanotubes science and applications. Boca Raton: CRC Press; 2005.

[11] Mesyats GA. Cathode phenomena in a vacuum discharge: the breakdown, the spark, and the arc. Moscow: Nauka Publ; 2000.

[12] Bloor D, Donnelly K, Hands PJ, Laughlin P, Lussey D. A metal-polymer composite with unusual properties. J Phys D Appl Phys 2005;38(16):2851–60.

[13] Han BG, Yu Y, Han BZ, Ou JP. Development of a wireless stress/strain measurement system integrated with pressure-sensitive nickel powder-filled cement-based sensors. Sens Actuators A Phys 2008;147(2):536–43.

[14] Han BG, Han BZ, Yu X, Ou JP. Ultrahigh pressure-sensitive effect induced by field emission at sharp nano-tips on the surface of spiky spherical nickel powders. Sens Lett 2011;9(5):1629–35.

[15] Han BG, Han BZ, Ou JP. Experimental study on use of nickel powder-filled Portland cement-based composite for fabrication of piezoresistive sensors with high sensitivity. Sens Actuators A Phys 2009;149(1):51–5.

[16] Han BG, Han BZ, Yu X. Experimental study on the contribution of the quantum tunneling effect to the improvement of the conductivity and piezoresistivity of a nickel powder-filled cement-based composite. Smart Mater Struct 2009;18. 065007(7pp).

[17] Han BG, Zhang K, Yu X, Kwon E, Ou JP. Nickel particle-based self-sensing pavement for vehicle detection. Measurement 2011;44(9):1645–50.

[18] Han BG, Han BZ, Yu X, Ou JP. Piezoresistive characteristic model of nickel/cement composites based on field emission effect and inter-particle separation. Sens Lett 2009;7(6):1044–50.

[19] Han BG, Han BZ, Yu X. Effects of content level and particle size of nickel powder on the piezoresistivity of cement-based composites/sensors. Smart Mater Struct 2010;19. 065012(6pp).

[20] Han BG, Guan XC, Ou JP. Electrode design, measuring method and data acquisition system of carbon fiber cement paste piezoresistive sensors. Sens Actuators A Phys 2007;135(2):360–9.

[21] French PJ, Evans AGR. Polycrystalline silicon as a strain gauge material. Solid-State Electron 1989;32:1–10.

[22] Chung DDL. Cement-matrix composites for smart structures. Smart Mater Struct 2000;9(4): 389–401.

[23] Han BG, Ou JP. Embedded piezoresistive cement-based stress/strain sensor. Sens Actuators A Phys 2007;138(2):294–8.

[24] Han BG, Chen W, Ou JP. Study on piezoresistivity of cement-based materials with acetylene carbon black. Acta Mater Compositae Sin 2008;25(3):39–44.

[25] Sun MQ, Li ZQ, Liu QP. The electromechanical effect of carbon fiber reinforced cement. Carbon 2000;40(12):2263–4.

[26] Wang YL, Zhao XH. Positive and negative pressure sensitivities of carbon fiber-reinforced cement-matrix composites and their mechanism. Acta Mater Compositae Sin 2005;22(4):40–6.

[27] Xue QZ. The influence of particle shape and size on electric conductivity of metal-polymer composites. Eur Polym J 2004;40:323–7.

[28] Feng JY, Chan CM. Effects of strain and temperature on the electrical properties of carbon black-filled alternating copolymer of ethylene-tetrafluoroethylene composites. Polym Eng Sci 2003;43(5):1064–70.

[29] Zallen R. The physics of amorphous solids. New York: Wiley; 1983.

[30] Zhou GD. Structures and properties. Beijing: Higher Education Press; 2000.

[31] Benamara Z, Mecirdi N, Akkal B, Mazari H, Chellali M, Talbi A, et al. Electrical characterization and electronic transport modelization in the InN/InP structures. Sens Lett 2009;7(5):712–5.

[32] Mao QZ, Zhao BY, Sheng DR, Li ZQ. Resistance changement of compression sensible cement speciment under different stresses. J Wuhan Univ Technol 1996;11(3):41–5.

[33] Geis MW, Efemow NN, krohn KE, Twichell JC, Lyszczarz TM, Kalish R, et al. A new surface electron-emission mechanism in diamond cathodes. Nature 1998;393:431–5.

[34] Chen K, Xiong CX, Li LB, Zhou L, Lei Y, Dong LJ. Conductive mechanism of antistatic poly (ethylene terephthalate)/ZnOw composites. Polym Compos 2008;30:226–31.

[35] Celzard A, Mcrae E, Furdin G, Mareche JF. Conduction mechanisms in some graphite-polymer composites: the effect of a direct-current electric field. J Phys Condens Matter 1997;9:2225–37.

[36] Azhari F. Cement-based sensors for structural health monitoring [dissertation for the Master Degree of Applied Science]. Vancouver, Canada: University of British Columbia; 2008.

[37] Huang Y, Xiang B, Ming XH, Fu XL, Ge YJ. Conductive mechanism research based on pressure-sensitive conductive composite material for flexible tactile sensing. In: Proc. of the 2008 IEEE international conference on information and automation (Zhangjiajie); 2008. pp. 1614–9.

[38] Sedlackova K, Lobotka P, Vavra I, Radnoczi G. Structural, electrical and magnetic properties of carbon-nickel composite thin films. Carbon 2005;43(10):2192–8.

[39] Lu JR, Chen XF, Lu W, Chen GH. The piezoresistive behaviors of polyethylene/foliated graphite nanocomposites. Eur Polym J 2006;42(5):1015–21.

[40] Rosner RB. Conductive materials for ESD applications: an overview. IEEE Trans Device Mater Reliab 2001;1(1):9–16.

[41] Zhang XW, Pan Y, Zheng Q, Yi XS. Time dependence of piezoresistance for the conductor-filled polymer composites. J Polym Sci B Polym Phys 2000;38(21):2739–49.

[42] Berdinsky AS, Shaporin AV, Yoo JB, Park JH, Alegaonkar PS, Han JH, et al. Field enhancement factor for an array of MWNTs in CNT paste. Appl Phys A Mater Sci Process 2006;83:377–83.

[43] Ahmad A, Tripathi VK. Model calculation of scanned field enhancement factor of CNTs. Nanotechnology 2006;17:3798–801.

[44] Pogorelov EG, Zhbanov AI, Chang YC. Field enhancement factor and field emission from a hemi-ellipsoidal metallic needle. Ultramicroscopy 2009;109:373–8.

[45] Lin YC, Bai JW, Huang Y. Self-aligned nanolithography in a nanogap. Nano Lett 2009;9:2234–8.

[46] Choi YC, Lee N. Influence of length distributions of carbon nanotubes on their field emission uniformity in the paste-printed dot arrays. Diam Relat Mater 2008;17:270–5.

[47] Tian F, Liu W, Wang CR. Controllable preparation of copper tetracyanoquinodimethane nanowire and the field emission study. J Phys Chem C 2008;112:8763–6.

[48] Xia LS, Zhang H, Yang XL, Pan HF, Liu YL. Experimental research on field emission of carbon fiber. High Power Laser Part Beams 2007;19:685–8.

[49] Guo ZH. Reinforced concrete theory and analyse. Beijing: Tsinghua University Press; 2004.

[50] Han BG, Yu X, Ou JP. Chapter 1: multifunctional and smart carbon nanotube reinforced cement-based materials. In: Gopalakrishnan K, Birgisson B, Taylor P, Attoh-Okine NO, editors. Nanotechnology in civil infrastructure: a paradigm shift. Springer; 2011. pp. 1–47. 276 pp.

[51] Yu Y. Wireless sensors and their network systems for structural health monitoring [dissertation for the Doctor Degree in Engineering]. Harbin Institute of Technology; 2006.

[52] Spencer BF, Ruiz-Sandoval ME, Kurata N. Smart sensing technology: opportunities and challenges. Struct Control Health Monit 2004;11:349–68.

[53] Wang DH, Liao WH. Wireless transmission for health monitoring of large structures. IEEE Trans Instrum Meas 2006;55(3):972–81.

Carbon-Nanotube-Based Self-Sensing Concrete

315

Self-Sensing Concrete in Smart Structures. http://dx.doi.org/10.1016/B978-0-12-800517-0.00010-1

10.1 Introduction and Synopsis

Carbon nanotube (CNT) is an allotrope of carbon with cylindrical nanostructure. It is generally a few nanometers in diameter and several micrometers in length. CNT is categorized as single-walled CNT (SWNT) and multi-walled CNT (MWNT) depending on the number of layers of atoms (as shown in Figure 10.1) [1–3]. In recent years, CNT has been widely applied in a variety of fields due to its excellent physical properties, such as high strength and Young's modulus (the tensile strength and Young's modulus of CNT are 20 times and 10 times that of carbon fiber (CF), respectively), high bonding force with matrix (the interlaminar shear strength of CNT reinforced epoxy materials is 10 times that of CF reinforced epoxy materials), large deformation and high ductility (the elongation at break of CNT is 18%, and is 18 times that of CF), high aspect ratio (>500), and excellent electrical conductivity. The extremely high aspect ratio, hollow structure, and low density of CNT makes it easy to form a conductive and mechanical reinforcement network inside concrete matrix with a CNT concentration level as low as 0.05 wt.%. CNT also has interesting piezoresistive properties. When CNT is subjected to stress/strain, its electrical properties will change with the level of stress/strain, expressing a linear and reversible piezoresistive response even for a large strain of 3.4%. In addition, the

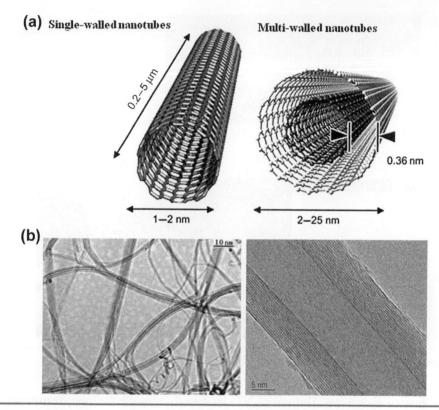

Figure 10.1 Carbon nanotube: (a) structure of carbon nanotube [1] (b) transmission electron microscope (TEM) photos of carbon nanotube [2,3].

small diameter and high aspect ratio of CNT is very favorable for motivating field-emission conduction. Because CNT is superior to CF in many respects and can be used as reinforcements to transfer the reinforcement behavior from the macroscale to the nanoscale level, it is promising to develop multifunctional or smart concrete without adding any additional weight and sacrificing other properties of concrete [4].

Li et al. first observed the pressure-sensitive responses of concrete with MWNT in 2007 [5]. Mohamed successfully fabricated self-sensing concrete by using SWNT as functional filler in 2009 [6]. Besides sensing properties, the concrete with CNT possesses excellent mechanical properties. The best observed performances include a 50% increase in compressive strength in a concrete with MWNTs [7], over 600% improvement in Vickers's hardness at early ages of hydration for a concrete with SWNTs [8], a 227% increase in Young's modulus for a concrete with MWNTs [9], and a 40% increase in flexural strength for a concrete with MWNTs [10]. The investigations of microstructure have indicated that if the nanotubes are well dispersed there may be potential for improving the mechanical properties of concrete materials

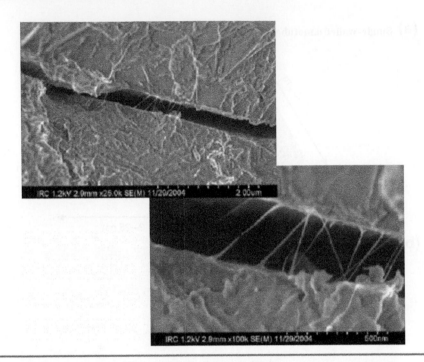

Figure 10.2 Concrete with carbon nanotube (CNT) [8].

in a more consistent way. The use of CNT can strongly reinforce the concrete materials at the nanoscale by increasing the amount of high-stiffness C-S-H and decreasing the porosity, which leads to the reduction of the autogenous shrinkage. Surprisingly, the addition of CNT also controls the formation of concrete-matrix cracks at the nanoscale level [10]. Evidence for crack bridging in concrete with MWNT has been observed (as shown in Figure 10.2). Other forms of classical reinforcing behaviors such as fiber-pullout and crack deflection have also been observed [8].

By now, much effort has been devoted to investigations on the CNT-based self-sensing concrete. In this chapter, we systematically introduce authors' research results on CNT-based self-sensing concrete, with attentions to its fabrication, measurement, self-sensing performances and mechanisms, and applications in traffic detection [11–17].

10.2 Fabrication of CNT-Based Self-Sensing Concrete

Since CNTs tend to self-associate into microscale aggregates, their disaggregation and uniform dispersion are critical challenges that must be addressed to successfully fabricate CNT concrete. Some studies have proved that the carboxyl CNT is much easier to disperse in water and concrete matrix than the plain CNT because of the carboxyl groups on the surface of CNT. Additionally, the carboxyl CNT has a better bond with concrete matrix than the plain CNT [18,19]. In addition to the carboxyl surface modification of CNTs, researchers have proved that sodium dodecylbenzene sulfonate (NaDDBS) and sodium dodecyl sulfate (SDS) can solubilize high weight fraction CNT in water [20,21]. Therefore, we choose carboxyl CNT and NaDDBS/ SDS to improve the dispersion of CNTs in concrete matrix, thus fabricating the self-sensing concrete. The materials and preparation of CNT concrete are as follows.

10.2.1 MATERIALS

The CNT used are MWNT provided by Timesnano, Chengdu Organic Chemicals Co. Ltd., China. Their properties are given in Table 10.1. The cement used is Portland cement (ASTM Type I) provided by Holcim Inc., USA. The sand used is commercial-grade fine sand provided by Quikrete International Inc., USA. The surfactants used for dispersing the MWNTs are NaDDBS and SDS, both provided by Sigma–Aldrich Co., USA. Tributyl phosphate (Sigma–Aldrich Co., USA) was used as defoamer to decrease the air bubble in CNT concrete caused by use of NaDDBS. Stainless steel gauzes with openings of 1.25×1.25 cm were used as embedded electrodes.

10.2.2 SPECIMEN PREPARATION

The mixing process of CNT concrete is illustrated in Figure 10.3. The NaDDBS or SDS was first mixed with water using a magnetism stirrer (PC-210, Corning Inc., USA) for 3 min. Next, carboxyl MWNTs were added into this aqueous solution and sonicated with an ultrasonicator (2510, Branson Ultrasonic Co., USA) for 3 h to make a uniformly dispersed suspension. Then, a mortar mixer was used to mix this suspension and cement or/and sand for about 3 min. Finally, defoamer was added into the mixture and mixed for another 1 min.

After the mixes were poured into oiled molds ($5.08 \times 5.08 \times 5.08$ cm) and two electrodes (5.08×5.08 cm) were embedded 1.5 cm apart, an electric vibrator was used to ensure good compaction. The specimens were then surface-smoothed, and covered with plastic films. All specimens were unmolded 24 h after casting.

TABLE 10.1 Properties of Carboxyl Multi-Walled CNT (MWNT)

Outside Diameter	Inside Diameter	-COOH Content	Length	Special Surface Area	Electrical Conductivity	Density
<8 nm	2~5 nm	3.86 wt.%	10–30 μm	>500 m^2/g	>10^2 s/cm	~2.1 g/cm^3

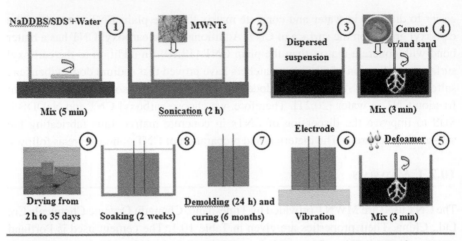

Figure 10.3   Fabrication process of self-sensing concrete with carbon nanotube (CNT).

Thereafter, they were cured under the standard condition at a temperature of 20 °C and a relative humidity of 100% for 28 days. All specimens were put at a room temperature for a period of time before they were tested.

10.3 Measurement of Sensing Signal of CNT-Based Self-Sensing Concrete

When measuring the sensing property of CNT concrete, the authors found that concrete composites have complex electrical properties, including resistance, capacitance, and impedance characteristics. This is because there are two basic types of electrical conduction in concrete composites: electronic conduction and ionic conduction. Electrons come from CNT, while ions come from hydrated cement [22–28].

The sensing property of CNT concrete results from electronic conduction, while ionic conduction will cause electrical polarization and affect sensing signal measurement of self-sensing concrete [22,29,30]. This chapter addresses this issue and develops effective measurement methods for sensing signals of the concrete under a polarization effect.

10.3.1 Experimental Description

The test setup in lab is illustrated in Figure 10.4. Resistance and capacitance of the specimens without loading were measured by using a handheld LCR meter (U1732A, Agilent Technologies, Inc., USA). A repeated compressive loading with stress amplitude of 6 MPa was applied using a material testing machine (ATS 900,

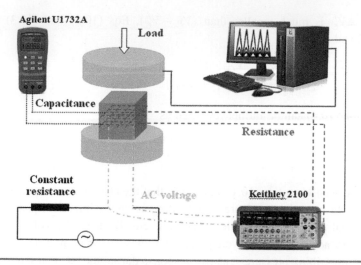

Figure 10.4 Sketch of loading test and three measurement circuits [11].

Applied Test Systems, Inc., USA). Under the compressive loading, resistance and capacitance of the specimens in compressive stress direction perpendicular to electrodes were measured by using a digital multimeter (Keithley Instruments Inc., USA) and the Agilent U1732A LCR meter, respectively.

When an AC measurement method is used, self-sensing concrete is series-connected with a constant reference resistance as shown in Figure 10.4, and we have

$$\frac{V'_C}{R'_C} = \frac{V_P - V'_C}{R_R} \qquad (10.1)$$

where R'_C and R_R are the resistance of the concrete and the constant reference resistance, respectively. V'_C is the measured voltage at both ends of the concrete. V_P is the voltage supplied by an external AC stabilized-voltage power.

Equation (10.2) can be further derived as

$$R'_C = \frac{V'_C}{V_P - V'_C} R_R \qquad (10.2)$$

If the composites are subjected to compressive loading, their resistance will vary. The change in resistance of the composites $\Delta R'_C$ can be expressed as

$$R'_C + \Delta R'_C = \frac{V'_C + \Delta V'_C}{V_P - \left(V'_C + \Delta V'_C\right)} R_R \qquad (10.3)$$

where $\Delta V'_C$ is the change in the voltage at both ends of the concrete.

Since $\Delta V_C'$ is much smaller than $(V_P - V_C')$, Eqs (10.2) and (10.3) can be combined and derived as

$$\frac{R_C' + \Delta R_C'}{R_C'} = \frac{V_C' + \Delta V_C'}{V_P - (V_C' + \Delta V_C')} \times \frac{V_P - V_C'}{V_C'} \approx \frac{V_C' + \Delta V_C'}{V_C'} \tag{10.4}$$

Equation (10.4) can be further rewritten as

$$\frac{\Delta R_C}{R_C} \approx \frac{\Delta V_C'}{V_C'} \tag{10.5}$$

It can be seen from Eqs (10.1) and (10.5) that the change in resistance signal of CNT concrete caused by compressive loading is approximately equal to that in voltage signal at both ends of the CNT concrete. Because the voltage signal is convenient to acquire, the AC voltage at both ends of the composites is taken as indices for describing pressure-sensitive response during the AC measurement. The voltage in circuit as shown in Figure 10.4 was provided by an AC power supply (15 V AC, Cui Inc., USA). The AC voltage signal at both ends of the CNT concrete was measured with a Keithley 2100 digital multimeter. All the experiments were conducted at room temperature.

10.3.2 RESISTANCE AND CAPACITANCE OF CNT-BASED SELF-SENSING CONCRETE WITHOUT COMPRESSIVE LOADING

Figure 10.5 shows the electrical resistivity and capacitance of CNT concrete under different test frequencies.

It can be seen from Figure 10.5(a) that the electrical resistivity of the specimens with different CNT concentrations decreases with the increase of test

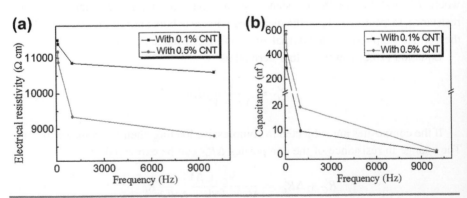

Figure 10.5   Electrical resistivity and capacitance of concrete with different carbon nanotube (CNT) concentrations [12]: (a) resistivity, (b) capacitance.

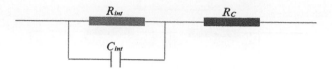

Figure 10.6   Conductive model of carbon nanotube (CNT) concrete [31], where R_C is the resistance of CNT in concrete and R_{int} and C_{int} are the interface resistance and the interface capacitance in concrete, respectively [11].

frequency. This can be attributed to the effect of the capacitance characteristic of CNT concrete as shown in Figure 10.5(b). Additionally, a higher test frequency leads to a lower capacitance, which is consistent with the conventional capacitive mechanism. The above results indicate that the CNT concrete has both resistance and capacitance characteristics. Furthermore, according to the conductive model of CNT concrete as shown in Figure 10.6, the concrete with 0.5% CNT has better CNT conductive network (i.e., lower R_C) and higher interface capacitance (i.e., higher C_{int}) than concrete with 0.1% CNT. As a result, the concrete with 0.5% CNT has lower electrical resistivity than that with 0.1% CNT, but it has higher capacitance.

Figure 10.7 depicts the variation of DC resistance with test time for CNT concrete without loading. As shown in Figure 10.7, the measured resistance increases rapidly during the initial stage of measurement. After that, the measured resistance first quickly and then slowly increases with test time. This phenomenon can be explained as below. The specimens with capacitance characteristic like capacitor, and will be charged during the measurement of resistance. This will generate a current opposite to the current within multimeter when a multimeter is

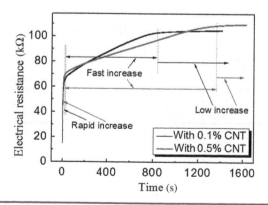

Figure 10.7   Variation of measured resistance with test time for carbon nanotube (CNT) concrete [11].

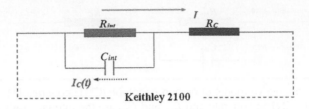

Figure 10.8   Equivalent circuit during DC measurement [11].

used to measure the resistance of specimens as shown in Figure 10.8. The resistance R_A' (i.e., apparent resistance, to be distinguished from the true resistance R_C') of specimens can be expressed as

$$R_A' = \frac{V}{I - I_C(t)} \tag{10.6}$$

where V and I are respectively voltage and current in the circuit within multimeter. $I_C(t)$ is the charging current of capacitor.

The charging of capacitor during the initial test stage is faster than that during the subsequent test stages, so the measured resistance exhibits a three-stage variation trend of rapid increase, fast increase, and low increase as shown in Figure 10.7.

10.3.3 RESPONSE OF CAPACITANCE OF CNT-BASED SELF-SENSING CONCRETE TO COMPRESSIVE LOADING

Figure 10.9 shows the variation of capacitance of CNT concrete under compressive loading. It can be seen from Figure 10.9 that the capacitance of the specimens

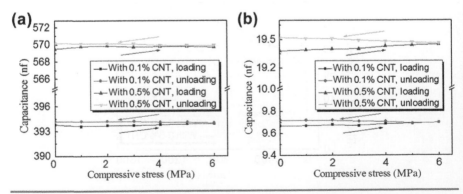

Figure 10.9   Response of capacitance of carbon nanotube (CNT) concrete to compressive loading [11]: (a) 100 Hz (b) 1000 Hz.

increases slightly during both loading and unloading under different test frequencies. There is not a regular corresponding relationship between compressive stress and capacitance of the specimens. This indicates that the compressive loading causes only a little insensitive and irregular change in capacitance.

10.3.4 RESPONSE OF RESISTANCE OF CNT-BASED SELF-SENSING CONCRETE TO COMPRESSIVE LOADING

Since the change in resistance trends to be stable in the "low increase" stage as shown in Figure 10.7, we measure the response of resistance of CNT concrete to compressive loading at this stage. The relationship curves between resistance (R'_A) and compressive stress (σ) are illustrated in Figure 10.10. This figure shows that the measured resistance of the specimens decreases upon loading and increases upon unloading in every cycle under the repeated compressive loading with amplitude of 6 MPa. However, the initial value of resistance in each loading and unloading cycle increases with the number of cycles. Because compressive loading hardly causes any change in capacitance of specimens, the compressive loading leads only to the reversible change of the true resistance of specimens (R'_C). In addition, as can be seen from Figure 10.10, the increase in measured resistance caused by the capacitor charging is approximately linear. Therefore, we remove the linear component in the measured resistance to obtain the sensing signals of the concrete. The resistance signals after removing the linear component are given in the two top traces in Figure 10.10. The two figures show that the change of resistance ($\Delta R'_C$) decreases reversibly during compressive loading and increases reversibly during unloading in every cycle. This further proves that the sensing signals of CNT concrete can be extracted by eliminating the linear increase component in the measured resistance.

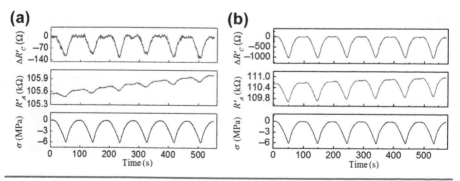

Figure 10.10   Response of resistance of self-sensing carbon nanotube (CNT) concrete to compressive loading [11]: (a) with 0.1% CNT, (b) with 0.5% CNT.

10.3.5 Response of AC Voltage of CNT-Based Self-Sensing Concrete to Compressive Loading

Figure 10.11 gives the response of AC voltage of self-sensing CNT concrete to compressive loading. As shown in this figure, the AC voltages (V_C') of the two kinds of specimens both present stable and reversible sensing property. This is because the capacitor allows the alternating current to pass as shown in Figure 10.12. The AC measurement method avoids the effect of capacitor charging and discharging on the sensing property of the concrete. Therefore, the AC voltage signals can be directly used to describe the sensing property of CNT concrete.

10.3.6 Effect of AC Voltage Amplitude on Sensing Property of CNT Concrete

When the self-sensing CNT concrete is series-connected to different values of constant resistance, the amplitude of the measured AC voltage at both ends of the concrete is different. Figure 10.13 shows the sensing property of specimens with 0.5% CNT under different amplitudes of measured AC voltage. It can be seen from Figure 10.13(a)–(e) that the change trends of sensing property under different voltage amplitudes are similar, but the sensing property curves are smoother and more regular at lower voltage amplitudes. This indicates that the smaller AC voltage signals have higher signal to noise ratio. In addition, it is noted from Figure 10.13(a)–(e) that the same compressive stress yields different levels of change in AC voltage signals. Figure 10.13(f) summarizes the fractional change in voltage at different amplitudes of AC voltages. This figure shows that the fractional change in voltage, i.e., the sensitivity of sensing property of specimens, increases with the decrease of AC voltage amplitude. This is because higher voltage amplitude at both ends of the concrete means that it is easier to induce capacitor charging and discharging during

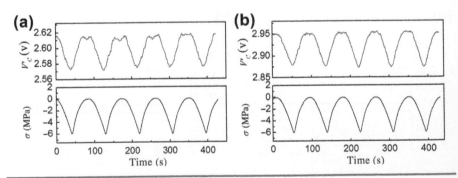

Figure 10.11   Response of AC voltage of self-sensing carbon nanotube (CNT) concrete to compressive loading [11]: (a) with 0.1% CNT, (b) with 0.5% CNT.

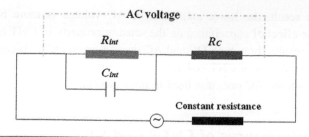

Figure 10.12   Equivalent circuit during AC measurement [11].

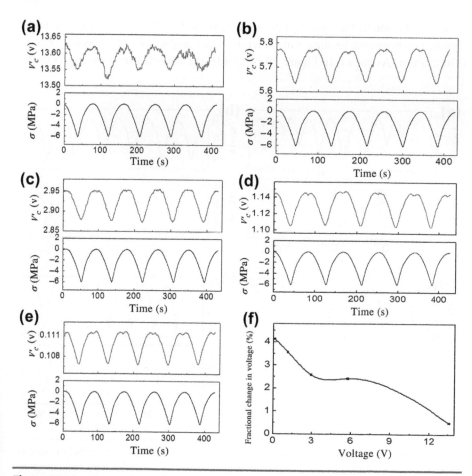

Figure 10.13   Comparison of sensing responses of carbon nanotube (CNT) concrete under different amplitudes of measured AC voltages [11]: (a) 13.615 V, (b) 5.772 V, (c) 2.950 V, (d) 1.145 V, (e) 0.111 V, and (f) change amplitude in voltage at 6 MPa of stress.

the test. As a result, the lower amplitude of AC voltage is more beneficial for eliminating the effect of capacitance on the sensing property of CNT concrete.

The above results show both DC and AC measurement methods can be used for the signal measurement of CNT concrete. Because the DC measurement method is easier to use than the AC one, it is used in the following study.

10.4 Performances of CNT-Based Self-Sensing Concrete

10.4.1 SENSING PERFORMANCE OF CNT-BASED SELF-SENSING CONCRETE AT DIFFERENT COMPRESSIVE STRESS AMPLITUDES

Figure 10.14 shows the sensing performance of self-sensing CNT concrete under repeated compressive loading with different stress amplitudes and a loading rate of 0.10 cm/min. As shown in Figure 10.14(a)–(d), the electrical resistance (R) of the

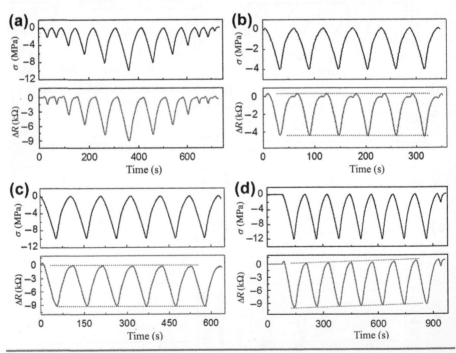

Figure 10.14 Sensing performance of self-sensing carbon nanotube (CNT) concrete under compressive loading with different amplitudes [12]: (a) under repeated compressive loading with different amplitudes, (b) under repeated compressive loading with amplitude of 4 MPa, (c) under repeated compressive loading with amplitude of 10 MPa, and (d) under repeated compressive loading with amplitude of 12 MPa.

CNT concrete decreases reversibly upon loading and increases reversibly upon unloading in every cycle under compressive loading. The higher compressive stress (σ) level yields bigger change in electrical resistance (ΔR, i.e.,$R - R_0$, where R_0 is the baseline electrical resistance of self-sensing CNT concrete and is defined as the maximum electrical resistance of the first loading and uploading cycle here). However, Figure 10.14 (d) shows that part of the ΔR appears irreversible when the compressive stress amplitude increases to 12 MPa; i.e., the initial value of ΔR in each loading and uploading cycle slightly increases with the cycle number.

The electrical resistance of self-sensing CNT concrete comes from two sources: the intrinsic resistance of nanotubes and the contact resistance at nanotube junctions (i.e., the resistance of the matrix connecting the crossing nanotubes and through which electrical tunneling occurs) as shown in Figure 10.15(a) and (b). Thus the electrical conductivity of the CNT concrete strongly depends on the morphology of the nanotube network and the number of contact points. The electric conductivity of individual CNT is about $10^4 \sim 10^7$ s/m. But the contact resistance is rather complicated and depends on nanotube diameter, tunneling gap at contact points, and matrix material filling the tunneling gap. The electrical resistance of self-sensing CNT concrete can be changed when it is deformed under applied loading. Several factors may contribute to the electrical resistance change. First, when self-sensing CNT concrete is deformed under external loading, the nanotube length and diameter will alter, resulting in the change of nanotube intrinsic resistance, and hence, the electrical resistance of the nanotube network. However, this resistance change is expected to be negligible because of the extremely small elastic deformation in nanotubes. The second and more important factor contributing to the resistance change of the sensors is the contact resistance. Under the applied load, the thickness of the insulating matrix between adjacent nanotubes may be changed considerably.

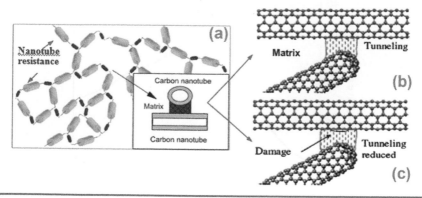

Figure 10.15 A schematic diagram of conductive network in carbon nanotube (CNT) concrete [32].

The compressive loading gives rise to the decrease of the gap at the contact area where electrical tunneling takes place and thus decreases the contact resistance [32]. Therefore, the sensing performance of CNT concrete results from the variation in the contact resistance among CNT and between CNT and concrete-matrix under compressive loading, which is caused by the deformation of CNT concrete.

Figure 10.16 gives the relationship between compressive stress and displacement of CNT concrete. As shown in this figure, the compressive strength of CNT concrete is about 40 MPa (a little higher than that of CNT concrete without electrodes because of the restriction effect of electrodes to composites), so the CNT concrete is still within an elastic regime when the stress amplitude is lower than 10 MPa. Since elastic deformation is recoverable, the sensing performance of the CNT concrete in the elastic regime is stable and reversible. However, when the compressive loading amplitude reaches 12 MPa, the deformation of CNT concrete possibly has gone beyond the elastic regime. Some minor damages appear inside the CNT concrete. Figure 10.15(c) illustrates the effect of damage on the change in contact resistance. The damaged spot assumes the form of a nanoscopic void, and its formation gives rise to the increase of the opening gap at the contact area where electrical tunneling takes place and thus increases the contact resistance [5]. As a result, the sensing performance of CNT concrete at compressive stress amplitude of 12 MPa is irreversible. The above experimental results are the same as that in references [33,34]. Therefore, the fabricated CNT concrete in this study can be used to achieve not only stress/strain monitoring by measuring the reversible resistance change, but also damage monitoring by measuring the irreversible resistance change.

In addition, CNT-based self-sensing concrete has higher signal to noise ratio than carbon-fiber-based self-sensing concrete [12,35]. The stable sensing responses of

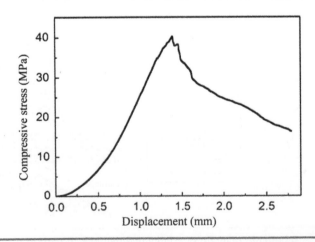

Figure 10.16 Relationship between compressive stress and displacement of carbon nanotube (CNT) concrete [12].

CNT concrete may result from the unique physical and chemical properties of CNT (e.g., ultra-high strength and stiffness, elastic stress-strain behavior, and excellent conductivity) and the extensive distributing conductive network of CNT in concrete [10,18,36].

10.4.2 SENSING PERFORMANCE OF CNT-BASED SELF-SENSING CONCRETE AT DIFFERENT LOADING RATES

Figure 10.17 depicts the sensing performance of CNT concrete at loading rates of 0.05, 0.10, 0.15, 0.20, 0.25, and 0.30 cm/min under repeated compressive load-ings with amplitude of 4 MPa. The sensing performance time-histories shown in Figure 10.17(a)–(d) indicate that the self-sensing CNT concrete present almost the same sensing responses at different loading rates. The relationship curves between ΔR and σ given in Figure 10.17(e) show that the loading rate almost has no effect on the sensing performance of the CNT concrete when it is below 0.20 cm/min. However, when the loading rate exceeds 0.20 cm/min, it has a little effect on the sensing performance of the CNT concrete, and the effect extent increases with the loading rate. This results from the sensing performance hysteresis of the CNT concrete to external loading, since the change in electrical resistance signals cannot catch up with that in external compressive loading when the loading rate increases. The higher the loading rate, the more obvious the sensing performance hysteresis.

10.4.3 SENSING PERFORMANCE PARAMETERS OF CNT CONCRETE

Since the above experimental results indicate that the fabricated self-sensing CNT concrete features stable and repeatable sensing performance when the compressive stress amplitude and the loading rate are lower than 10 MPa and 0.20 cm/min, respectively, this section investigates the sensing performance of the CNT concrete with compressive stress under 10 MPa and a loading rate of 0.10 cm/min.

Figure 10.18 shows the relationship between ΔR and σ. It can be seen from this figure that the relationship curve between ΔR and σ has good linearity when the compressive stress is lower than 10 MPa. The relationship between ΔR and σ can be expressed as

$$\Delta R = -0.911\sigma \qquad (10.7)$$

And such parameters of sensors as input, output, sensitivity, linearity, repeat-ability, and hysteresis are tabulated in Table 10.2. The concrete with 10% conductive CF fillers reported in reference [33] achieve a sensitivity of 1.35% (MPa), a linearity of 4.17%, a repeatability of 4.05%, and a hysteresis of 3.61% when the compressive stress varies from 0 to 8 MPa. And the concrete with 24% of conductive nickel

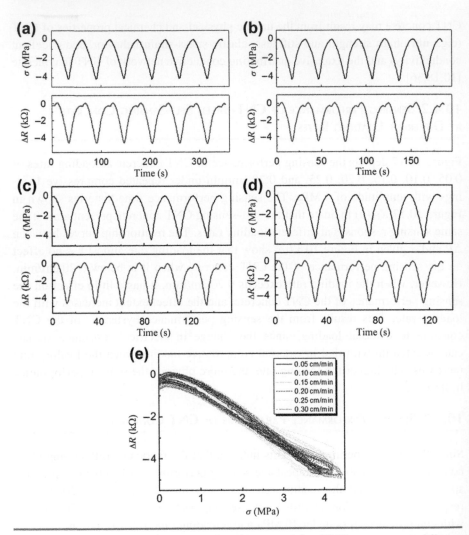

Figure 10.17 Sensing performance of carbon nanotube (CNT) concrete at different loading rates [12]: (a) loading rate of 0.10 cm/min, (b) loading rate of 0.20 cm/min, (c) loading rate of 0.25 cm/min, (d) loading rate of 0.30 cm/min and (e) comparison of piezoresistive responses of sensors with different loading rates.

powder fillers reported in reference [37] has a sensitivity to compressive stress higher than 5% (MPa) when the compressive stress varies from 0 to 12.5 MPa. Here it can be seen from Table 10.2 that the repeatability of CNT concrete is lower than that of the sensors in references [33,37]. This indicates that the CNT concrete has more stable sensing properties. Although the other performance parameters of the CNT concrete are not good as that of the sensors in references [33,37], the content

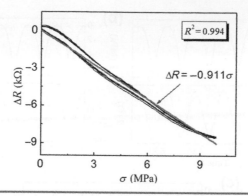

Figure 10.18 Relationship between ΔR and σ [12].

TABLE 10.2 Performance Parameters of Carbon Nanotube (CNT) Concrete

Performance Parameters	Values
Input	0–10 MPa
Output	0–9.81 kΩ
Sensitivity	0.911 kΩ/MPa
Linearity	7.16%
Repeatability	1.53%
Hysteresis	7.24%

level of conductive fillers adopted in this paper is only about 0.1%, which is much lower than that of conductive fillers adopted in references [33,37]. In addition, unlike microscale conductive fibers and particles, CNT has a significant enhancement to microstructure of concrete materials at nanoscale. It is expected that the fabricated CNT concrete in this study is superior in mechanical properties and long-term durability to other concrete composites. Furthermore, the gauge factor of CNT concrete is about 50, which is about 20 times the gauge factor of a conventional resistive strain gauge.

10.4.4 EFFECTS OF CNT CONCENTRATION LEVEL ON THE SENSING PROPERTY

Figure 10.19 shows the sensing responses under repeated compressive loading with amplitude of 6 MPa for concrete with 0.05%, 0.1%, and 1% of CNT. As can be seen in this figure, the electrical resistance R of all the three types of CNT concrete

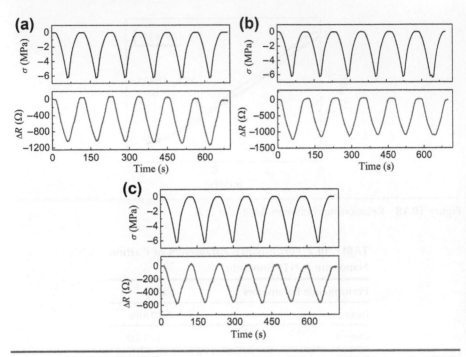

Figure 10.19 Sensing property of concrete with different carbon nanotube (CNT) concentration levels [13]: (a) with 0.05 wt.% of CNT, (b) with 0.1 wt.% of CNT, and (c) with 1 wt.% of CNT.

decreases upon loading and increases upon unloading in every cycle under compressive loading, expressing stable and regular sensing property.

Figure 10.20 depicts the change amplitudes of the electrical resistance for concrete with 0.05%, 0.1%, and 1% of CNT as the compressive stress is 6 MPa. As shown in Figure 10.20, we find that the change in electrical resistance ΔR (i.e., $R - R_0$, where R_0 is the initial electrical resistance of samples without compressive loading) of concrete with 0.05%, 0.1%, and 1% of CNT reaches about 1000, 1150, and 600 Ω, respectively, as compressive stress is 6 MPa. This indicates that the CNT concrete with 0.1 wt.% of CNT has the most sensitive response to compressive loading among the three types of CNT concrete. It is interesting to note that the sensing sensitivity does not linearly increase with CNT concentration levels. This phenomenon can be explained below. The concentration level of CNT influences and reflects the situation of network formed in the composites. When the CNT concentration level is 0.05 wt.%, the thickness of the insulating matrix between adjacent nanotubes is large, and the amount of conducted tunneling junction under external loading is low. With the increase of the concentration level of CNT to 0.1 wt.%, the thickness of the insulating matrix between adjacent nanotubes decreases and the electronic transition by tunneling conduction becomes easy. With

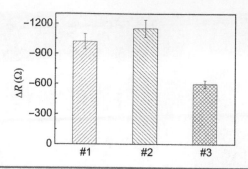

Figure 10.20 Comparison of electrical resistance changes of concrete with different carbon nanotube (CNT) concentration levels (#1: 0.05 wt.%, #2: 0.1 wt.%, #3: 1 wt.%) [13].

the continuous increase of CNT, the tunneling gap would be further shortened, and then the CNT network stabilizes and becomes hardly to change under loading. This can be proved by SEM photographs as shown in Figure 10.21. Comparing Figure 10.21(b) with Figure 10.21(a), it can be found that the CNT network in CNT concrete with 1 wt.% of CNT is more widespread than that in CNT concrete with 0.1 wt.% of CNT. As a result, the contact resistance of CNT concrete with 0.1 wt.% of CNT is the most sensitive to compressive loading among the fabricated CNT concretes. Therefore, the sensitivity of sensing response of CNT concrete first increases then decreases with the increase of CNT concentration levels [38–41].

10.4.5 Effects of Water/Cement Ratio on the Sensing Property

As seen from Figure 10.22(a), Figure 10.22(b), and Figure 10.23, it is noted that the change in electrical resistance of CNT concrete with 0.6 water/cement ratio and NaDDBS as surfactant decreases about 1500 Ω as the compressive stress is 6 MPa, which is obviously larger than 1150 Ω of CNT concrete with 0.4 water/cement ratio and NaDDBS as surfactant. In addition, Figure 10.22(c) shows the results of similar CNT concrete to Figure 10.22(b), except that the surfactant used is SDS. The trend for the variation in the electrical resistance under loading is almost the same as in Figure 10.22(b). The above results indicate that the increase of water/cement ratio improves the sensitivity of sensing responses no matter whether NaDDBS or SDS is used as surfactant. Two factors would contribute to the effect of water/cement ratio on the sensitivity of sensing response. One is the deformation capacity of CNT concrete, and the other is the dispersion of CNT in concrete composites. Figure 10.24 gives the relationships between compressive stress and displacement of CNT concrete with 0.4 water/cement ratio and NaDDBS as surfactant and CNT concrete with 0.6 water/cement ratio and NaDDBS as surfactant. As shown in this

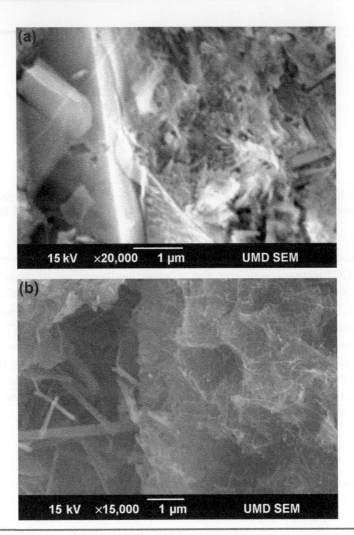

Figure 10.21 SEM photographs of carbon nanotube (CNT) concrete [13]: (a) with 0.1 wt.% of CNT, (b) with 1 wt.% of CNT.

figure, the CNT concrete with 0.4 water/cement ratio and NaDDBS as surfactant has bigger deformation than CNT concrete with 0.6 water/cement ratio and NaDDBS as surfactant at same compressive stress. The conductive network in concrete with a higher water/cement ratio is thus easier to change and the electrical resistance of such composite is more sensitive to loading. In addition, with a higher water/cement ratio, the CNT/water ratio is lower which helps the dispersion of CNT in CNT concrete [20,21,39]. Therefore, the electrical resistance of CNT concrete with higher water/cement ratio is more sensitive to compressive loading. Additionally,

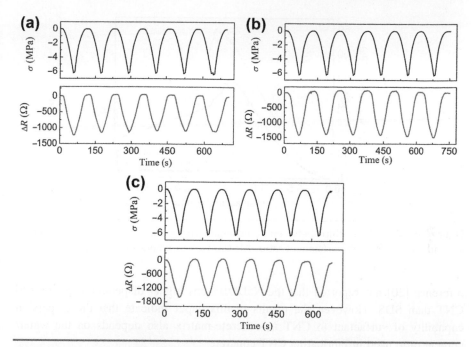

Figure 10.22 Sensing property of carbon nanotube (CNT) concrete with different water/cement ratios [13]: (a) with 0.4 water/cement ratio and NaDDBS as surfactant (sample #2), (b) with 0.6 water/cement ratio and NaDDBS as surfactant (sample #4), and (c) with 0.6 water/cement ratio and SDS as surfactant (sample #5).

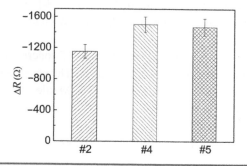

Figure 10.23 Comparison of electrical resistance changes of carbon nanotube (CNT) concrete with different water/cement ratios (#2: 0.45 water/cement ratio and NaDDBS as surfactant, #4: 0.6 water/cement ratio and NaDDBS as surfactant, #4: 0.6 water/cement ratio and SDS as surfactant) [13].

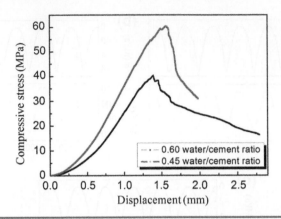

Figure 10.24 Relationships between compressive stress and displacement of samples #2 and #4 (#2: 0.45 water/cement ratio, #4: 0.6 water/cement ratio) [13].

reference [20] has reported that the NaDDBS has higher dispersion capability to CNT than SDS. However, the results in this paper indicate that the dispersion capability of surfactant to CNT in concrete-matrix also depends on the water/cement ratio used in fabricating CNT concrete.

10.4.6 EFFECT OF WATER CONTENT ON THE SENSING PROPERTY

Water is a main raw material for fabrication of concrete, and the water content in the concrete composites fluctuates under internal hydration and external environmental effects [41,42]. In addition, the adsorption of water gives rise to the change in the intrinsic resistance of nanotubes [43–46]. Especially, the water attracted to the nanotube tip makes the highly occupied molecular orbital unstable, thereby enhancing the field emission on the nanotube tip [47–50]. Therefore, the effect of water in CNT concrete on the sensing property needs to be investigated for real road/civil applications.

The water content and the sensing property of CNT concrete were measured over a drying period of time ranging from 2 h to 35 days. The water content measurement was measured using an electronic scale (PG5002-S, Mettler Toledo Inc., USA). The water content C_w can be denoted by

$$C_w = \frac{W_s^t - W_s^0}{W_s^0} \times 100\% \tag{10.8}$$

where W_s^t is the mass of CNT concrete when the time is t during drying, W_s^0 is the constant mass of CNT concrete when the drying time exceeds 28 days under a temperature of 20 °C and a relative humidity of 30%.

Figure 10.25 shows the sensing property of CNT concrete with different water contents under repeated compressive loading with amplitude of 6 MPa. The sensitivities k ($k = \left| \frac{\Delta R}{\Delta \sigma} \right| \times 100\%$) are fitted using a linear regression. As can be seen in this figure, the electrical resistance of CNT concrete decreases linearly and reversibly upon loading and increases linearly and reversibly upon unloading in every cycle under compressive loading, expressing regular and stable sensing property, while composites with different water contents yield different levels of resistance changes.

Figure 10.26 shows the initial electrical resistances R_0, the maximum change amplitudes f ($f = \left| \frac{R_{6 \text{ MPa}} - R_0}{R_0} \times 100\% \right|$), where $R_{6 \text{ MPa}}$ is the electrical resistance of CNT concrete when the compressive stress is 6 MPa) of electrical resistance and sensitivities of CNT concrete with different water contents. It can be found from Figure 10.26 that the maximum change amplitudes of electrical resistance and sensitivity of CNT concrete with 3.3% of water content are the highest among composites with different water contents. The above results indicate that the sensitivities of the composites first increase and then decrease with the increase of water content in the composites.

It is interesting to note that the sensitivity of the CNT concrete does not linearly increase with water content, but the electrical conductivity of the composites as shown in Figure 10.26 increases with water content. This phenomenon can be explained below. Two factors would contribute to the effect of water content on the sensitivity of sensing property. One is the electrical conductivity of matrix [51], and the other is the field-emission effect on the nanotube tip. The electrical conductivity of matrix and the field-emission effect on the nanotube tip can be enhanced by the adsorption of water molecules [5,47–49]. When the water content is 0.1%, the electrical conductivity of matrix filling the tunneling gap is low (i.e., the contact resistance is high) and the field-emission effect on the nanotube tip is weak. The conductive path is thus hard to form, even when an external force is applied to the composites. As a result, the composites possess high electrical resistance and low sensitivity to stress. With the increase of the water content to 3.3%, the electrical conductivity of matrix filling the tunneling gap increases (i.e., the contact resistance decreases) and the field-emission effect on the nanotube tip is enhanced. This increases the electrical conductivity of composites. Furthermore, when the composites deform under compressive loading, the tunneling barrier of electrons will decrease and the field-emission-induced tunneling can easily occur in the composites. These cause the concrete to present lower electrical resistance and higher sensitivity. With the continuous increase of water content to a higher level such as 9.9%, the electrical conductivity of matrix filling the tunneling gap further increases (i.e., the contact resistance further decreases) and the field-emission effect on the nanotube tip is enhanced, and then the conductive network stabilizes and becomes hard to change

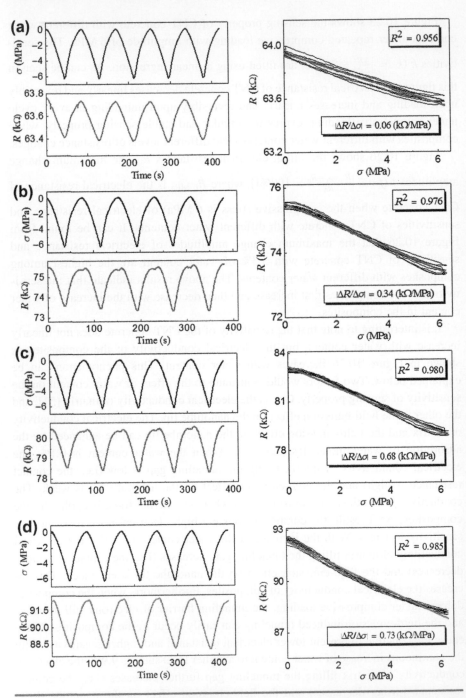

Figure 10.25 Sensing property of carbon nanotube (CNT) concrete with different water contents [14]: (a) with 9.9% water content, (b) with 7.6% water content, (c) with 5.7% water content, (d) with 3.3% water content, (e) with 1.3% water content, and (f) with 0.1% water content.

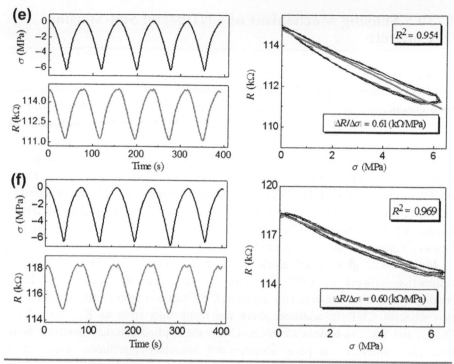

(e)

$R^2 = 0.954$

$\Delta R/\Delta\sigma = 0.61\,(\text{k}\Omega/\text{MPa})$

(f)

$R^2 = 0.969$

$\Delta R/\Delta\sigma = 0.60\,(\text{k}\Omega/\text{MPa})$

Figure 10.25   *(continued)*

under loading. As a result, too high a water content will induce a much lower electrical resistance and a lower sensitivity to stress [34,50–53]. Therefore, the sensitivity of self-sensing CNT concrete is highly influenced by the water content in the concrete, and it first increases then decreases with the increase of water content in the concrete.

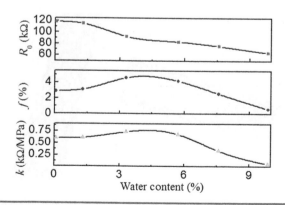

Figure 10.26   Comparison of electrical resistances, maximum change amplitudes of electrical resistance, and sensitivities of carbon nanotube (CNT) concrete with different water contents [14].

10.5 Sensing Mechanism of CNT-Based Self-Sensing Concrete

Previous researchers considered that the sensing property of self-sensing concrete with CNT is caused by the following four reasons: (1) the change in intrinsic electrical conductivity of CNT varies under external force; (2) the number of contact points of CNT increases with the increase of loading, which can cause an enhancement of conductivity; (3) the separation distance between adjacent CNT decreases under compressive loading, which can cause an enhancement in tunneling effect conduction; (4) the field-induced tunneling effect enhances due to the material's deformation making the CNT closer. All the four reasons can be summarized as two aspects, i.e., the change in intrinsic resistance of CNT and the change in contact resistance between adjacent CNT [4,5]. However, it is not clear about how the two factors contribute to the sensing property of the CNT concrete.

Figure 10.27 gives a schematic diagram of a basic conductive element in conductive network of CNT concrete. As shown in this figure, the basic conductive element contains two adjacent CNT and concrete matrix between the two adjacent CNT. In addition, there are many ions such as K^+, Na^+, Ca^+, OH^{--}, and SO_4^{2-} in hydrated cement. Under the applied external electrical field, ions are "blocked" at pore water/crystal boundary interfaces, while other charges, electrostatically held onto grain/crystal surfaces, can oscillate or diffuse (short-range) in sympathy with the applied alternating field. Such changes contribute to the polarization process. Therefore, the conductive model of the basic conductive element can be illustrated as Figure 10.28. It can be seen from this figure that the basic conductive element is an impedance element Z^i, which can be expressed as.

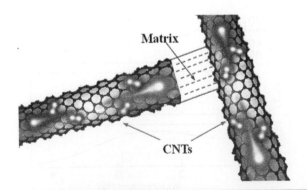

Figure 10.27 Schematic diagram of basic conductive element in conductive network of carbon nanotube (CNT) concrete [15].

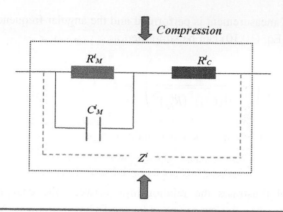

Figure 10.28   Conductive model of basic conductive element in conductive network of carbon nanotube (CNT) concrete, where R^i_C is the resistance of CNT in the ith basic conductive element, and R^i_M and C^i_M are the contact resistance and capacitance of matrix between adjacent CNT in the ith basic conductive element, respectively [15,54,55].

$$Z^i = R^i_C + \frac{R^i_M}{\sqrt{1 + \omega^2 (C^i_M)^2 (R^i_M)^2}} \tag{10.9}$$

where ω is angular frequency.

The total impedance of the composite can then be described by

$$Z = \sum_{i=1}^{n} \left(R^i_C + \frac{R^i_M}{\sqrt{1 + \omega^2 (C^i_M)^2 (R^i_M)^2}} \right) = R_C + \sum_{i=1}^{n} \left(\frac{R^i_M}{\sqrt{1 + \omega^2 (C^i_M)^2 (R^i_M)^2}} \right) \tag{10.10}$$

where n is the number of conductive elements in the whole composite sample, and R_C is the resistance sum of CNT that contributes to conductive network of the composite sample.

When a DC measurement is performed, ω is equal to zero. The Eq. (10.10) can be expressed as

$$Z = R_C + R_M \tag{10.11}$$

where R_M is the sum of contact resistance between adjacent CNT in the composite sample.

If a compressive stress is applied to the sample, the impedance will be altered due to the deformation of both CNT and composite matrix. The change in impedance of the composite under compression can be given by

$$\Delta Z = \Delta (R_C + R_M) \tag{10.12}$$

When an AC measurement is performed and the angular frequency ω of the AC voltage is high, Eq. (10.10) can be simplified as

$$Z \approx R_C + \sum_{i=1}^{n} \left(\frac{R_M^i}{\sqrt{\omega^2 \left(C_M^i\right)^2 \left(R_M^i\right)^2}} \right) = R_C + \sum_{i=1}^{n} \left(\frac{1}{\omega C_M^i} \right) \approx R_C \qquad (10.13)$$

And then, the change in impedance under compression can be calculated as

$$\Delta Z \approx \Delta R_C \qquad (10.14)$$

Figure 10.29 illustrates the relationships between the change in resistance $(R_C + R_M)$ and compressive stress (σ) under the DC measurement. As shown in this figure, the $(R_C + R_M)$ decreases upon loading and increases upon unloading in every cycle under the repeated compression, and its change reaches about 200 Ω as the compressive stress is 6 MPa. Figure 10.30 depicts the relationships between the change in resistance R_C and σ under the AC measurement. This figure shows that the variation trend of R_C under the loading is the same as that of $(R_C + R_M)$ in Figure 10.29. However, the change amplitude of R_C is only about 5 Ω as the compressive stress is 6 MPa, much lower than the change amplitude in $(R_C + R_M)$ under the same stress amplitude. This indicates that both the change in resistance of CNT and the change in contact resistance between adjacent CNT have contribution to sensing property of the concrete with CNT under compression. However, the contribution of the change in resistance of CNT to sensing property almost can be neglected compared with that of the change in contact resistance. Tombler et al. [56] and Pushparaj et al. [57] have reported that the deformation of CNT can cause

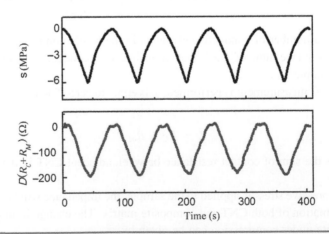

Figure 10.29 Sensing property of carbon nanotube (CNT) concrete with DC measurement [15].

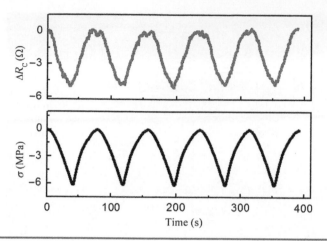

Figure 10.30 Sensing property of carbon nanotube (CNT) concrete with AC measurement [15].

remarkable response of their intrinsic electrical resistance. However, CNT have much higher elastic modulus than concrete matrix and are supported by rigid concrete matrix. Under the compressive loading, CNT do not deform much. Therefore, the change in the intrinsic electrical resistance of CNT performs as a smaller contributing factor for sensing property of the CNT concrete.

10.6 Application of CNT-Based Self-Sensing Concrete in Traffic Detection

10.6.1 Verification of Feasibility of Using Self-Sensing Concrete for Traffic Detection

Figure 10.31 shows the road test setting. Self-sensing concrete is embedded in concrete pavement as shown in Figure 10.31(a). Vehicles are driven over the concrete to investigate the feasibility of traffic detection, as shown in Figure 10.31(b).

The variation in electrical resistance of self-sensing CNT concrete under vehicular loading is illustrated in Figure 10.32, where Figure 10.32(a) shows the changes of resistance when two midsize passenger vehicles pass over the sensor and Figure 10.32(b) shows the results when a minivan vehicle passes over the sensor. As can be seen in Figure 10.32, vehicular loads can lead to remarkable change in electrical resistance of the sensors. Comparing Figure 10.32(b) with Figure 10.32(a) carefully, we can also find that the change amplitude in electrical resistance of minivan passing over is larger than that of passenger vehicles, which is due to the heavier axis weight of the minivan (i.e., larger mechanical stress). These findings

Figure 10.31 Vehicular loading experiment of self-sensing concrete embedded in concrete pavement [16]: (a) self-sensing carbon nanotube (CNT) concrete sensor embedded in concrete pavement, and (b) vehicular loading experiment.

indicate that self-sensing CNT concrete can detect traffic flow and even possible to identify different vehicular loadings (weight-in-motion detection). The weight of vehicles such as heavy trucks is needed to avoid damage to highways due to over-weight. It is currently being examined in weighing stations off the highway while the vehicle is stationary. The monitoring of the weight of vehicles can be more convenient and effective if the weighing is performed on the highway while the vehicle is moving normally. In this way, traffic is not affected and time is saved.

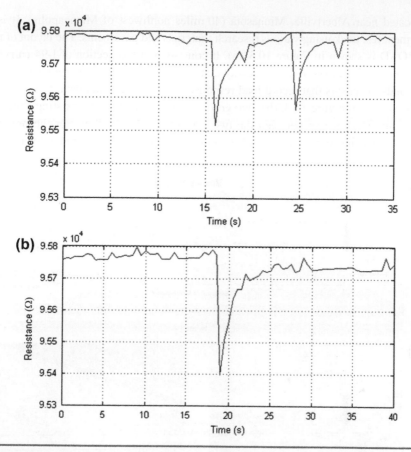

Figure 10.32   Variation in electrical resistance of the self-sensing carbon nanotube (CNT) concrete under different vehicular loadings [16]: (a) two middle-size passenger vehicles pass over the sensor, and (b) a minivan passes over the sensor.

According to the results and analyses, the self-sensing concrete pavement embedded with self-sensing CNT concrete presents great potential for traffic monitoring such as traffic flow detection, weigh-in-motion detection, and vehicle-speed detection.

10.6.2 INTEGRATION AND ROAD TESTS OF A SELF-SENSING CONCRETE PAVEMENT SYSTEM FOR TRAFFIC DETECTION

10.6.2.1 Construction of Self-Sensing Concrete Pavement

Two CNT-based self-sensing concrete sensors, a pre-cast sensor and a cast-in-place sensor, were integrated into a concrete test section at the Minnesota Road Research Facility (MnROAD) of the Minnesota Department of Transportation, USA. MnROAD is located near Albertville, Minnesota (40 miles northwest of Minneapolis). It is a pavement test track using various research materials and pavements. The layout of the MnROAD is shown in Figure 10.33(a). It consists of a test section of I-94 carrying interstate traffic, a low-volume roadway that simulates conditions on rural roads, and thousands of sensors that record load response and environmental data.

As shown in Figure 10.33(b), two grooves were first cut in the existing concrete pavement. The spacing between the two grooves is about 1.8 m. For the pre-cast CNT self-sensing concrete, the CNT concrete mixture was poured into a wood

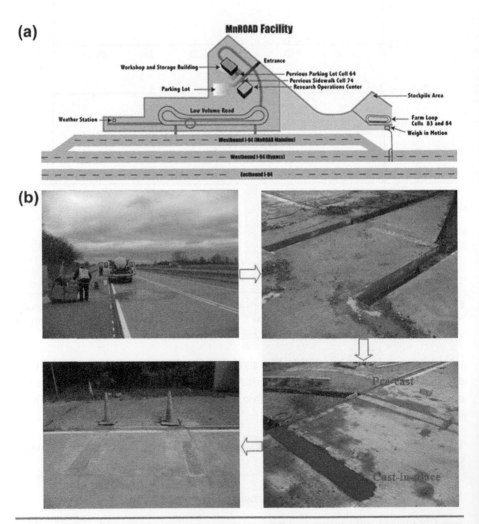

Figure 10.33 Location and construction of self-sensing carbon nanotube (CNT) concrete pavement [17]: (a) layout of Minnesota Road Research Facility (MnROAD) testing facility, and (b) construction process of self-sensing CNT concrete pavement.

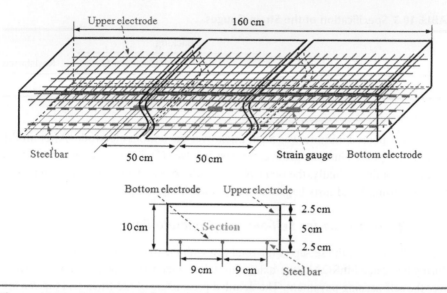

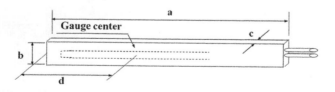

Figure 10.34 Structure of the carbon nanotube (CNT)-based self-sensing concrete sensors [17].

mold of $160 \times 23 \times 10$ cm with three reinforcing steel bars, three strain gauges (PML-60-2LT, Tokyo Sokki Kenkyujo Co., Ltd, Japan), and two mesh electrodes arranged as shown in Figure 10.34. The structure and specification of the strain gauges are given in Figure 10.35 and Table 10.3. A vibration table was used to ensure good compaction. The CNT concrete sensor was then surface-smoothed, and covered with plastic film to prevent water evaporation. After that, the sensor was cured at room temperature for 28 days before being installed in the road pavement. Finally, the sensor was fixed in one of the cutting grooves using concrete mortar. For the cast-in-place sensor, common patch mix concrete was first poured into the bottom of the other cutting groove until level with the bottom of the pre-cast CNT concrete sensor. Then, the CNT concrete mixture was poured into the groove, in which three reinforcing steel bars, three strain gauges, and the bottom mesh electrode were preinstalled as shown in Figure 10.34. The upper mesh electrode was put

Figure 10.35 Structure of the strain gauges [17].

TABLE 10.3 Specification of the Strain Gauges

Gauge Length	Gauge Width	Backing				Resistance
		A	B	C	D	
60 mm	1 mm	125 mm	13 mm	5 mm	40 mm	120 Ω

in when the poured mixture was 5 cm thick over the bottom electrode. After all the mixture was poured into the groove, the cast-in-place CNT concrete sensor was then surface-smoothed. Finally, the two CNT concrete sensors were covered with plastic film for curing. Road tests began after one month of curing.

10.6.2.2 Road Tests for Self-Sensing Concrete Pavement

Figure 10.36 shows the set-up of the road test with the road-side data collection unit. During the test, a MnROAD five-axle semi trailer-tractor truck and a van were driven over the self-sensing pavement. The detailed parameters of the truck and the van are given in Figure 10.37 and Table 10.4, respectively.

The measurement circuit diagram of the CNT-based self-sensing concrete sensors and strain gauges is depicted in Figure 10.38. The change in electrical resistivity ρ_S of the self-sensing concrete sensors is the same as that in electrical resistance R_S i.e.,

$$\Delta\rho_S/\rho_S = \Delta R_S/R_S \tag{10.15}$$

As shown in Figure 10.39, the CNT concrete sensors are series-connected with a constant reference resistance, and so we have

$$\frac{U_S}{R_S} = \frac{U_C - U_S}{R_R} \tag{10.16}$$

Figure 10.36 Road test of self-sensing carbon nanotube (CNT) concrete pavement [17].

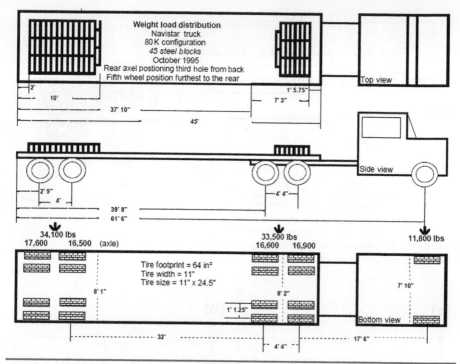

Figure 10.37 The MnROAD five-axle semi tractor-trailer truck [17].

where R_S and R_R are the resistance of the CNT concrete sensors and the reference resistance, respectively. U_S is the voltage at both ends of the CNT concrete sensors. U_C is the external constant voltage.

Equation (10.16) can be further derived as

$$R_S = \frac{U_S}{U_C - U_S} R_R \qquad (10.17)$$

Since U_C and R_R are known constants, it can be seen from Eq. (10.17) that the electrical resistance of the CNT concrete sensors is a one-variable function of the voltage at both ends of the CNT concrete sensors.

In addition, the sensing signals of the strain gauges can be collected by using the common Wheatstone bridge method as shown in Figure 10.38. The voltages at both

TABLE 10.4 Parameters of the Van

Van Model	Weight	Wheelbase
1999 Ford Econoline van E−250	~2500 kg	3.5 m

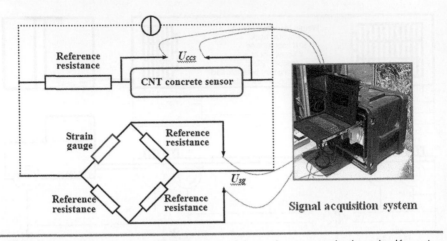

Figure 10.38 Measurement circuit diagram of the carbon nanotube-based self-sensing concrete sensors and strain gauges [17].

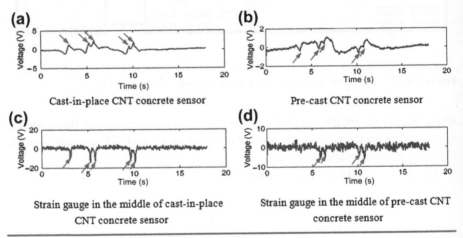

Figure 10.39 Detection results of truck passing at low speed (5 MPH, 8 km/h) [17], (a) Cast-in-place CNT concrete sensor, (b) Pre-cast CNT concrete sensor, (c) Strain gauge in the middle of cast-in-place CNT concrete sensor, (d) Strain gauge in the middle of pre-cast CNT concrete sensor.

ends of the CNT concrete sensors and the strain gauges were taken as indices for detecting the passing vehicles, since the electrical resistance of both the CNT concrete sensors and the strain gauges would change when the vehicles pass. As shown in Figure 10.36, a MnROAD signal acquisition system was used to collect the sensing signals of the two CNT concrete sensors and the six strain gauges. The sampling rate of the voltage signals is 1000 Hz. In addition, in order to decrease the effect of measurement noise, a low-pass filter was used to post-process the measured sensing signals of the CNT concrete sensors and the strain gauges.

10.6.2.3 Detection of Truck Passing

Detection results of truck passing at low and higher speeds are illustrated in Figure 10.39 and 10.40, respectively. As shown in Figure 10.39(a) and (b), Figure 10.40(a) and (b), abrupt changes occurred in the voltage signal curves when the truck passes over the cast-in-place and pre-cast CNT concrete sensors. Each peak indicates a passing wheel, which corresponds to the structure of the truck as shown in Figure 10.37. In addition, because truck wheels pass over the middle region of the CNT concrete sensors during test, only strain gauges in the middle of the two concrete sensors have some responses to the truck passing (the voltage signal curves of the strain gauges are also given in Figures 10.39 and 10.40). A comparison between the voltage signals of the CNT concrete sensors and those of the strain gauges indicates that when the truck passes, the detection results of the CNT concrete sensors are coincident with that of strain gauges. Additionally, the CNT concrete sensors generally have higher detection accuracy than the strain gauges (as shown in Figures 10.39(d) and 10.40(d), there were missed measurements on the strain gauge). This owes to the larger sensing area of the CNT pavement sensors.

Figure 10.41 illustrates the truck passing detection results in another two tests of the cast-in-place CNT concrete sensor and the strain gauge. The detection results shown in Figure 10.41 are very similar to those shown in Figures 10.39 and 10.40.

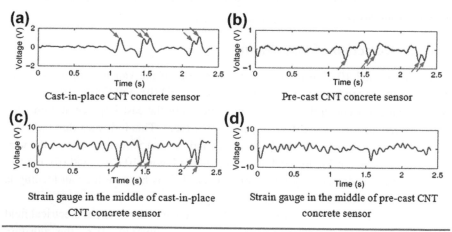

Figure 10.40 Detection results of truck passing at higher speed (20 MPH, 32 km/h) [17], (a) Cast-in-place CNT concrete sensor, (b) Pre-cast CNT concrete sensor, (c) Strain gauge in the middle of cast-in-place CNT concrete sensor, (d) Strain gauge in the middle of pre-cast CNT concrete sensor.

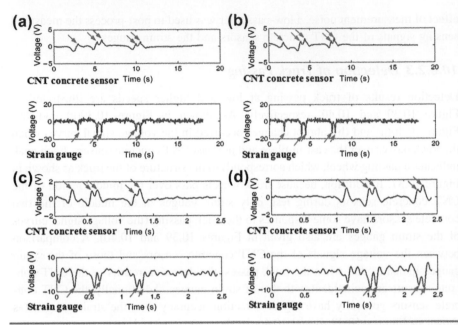

Figure 10.41 Detection results of cast-in-place carbon nanotube (CNT) concrete sensor and the strain gauge in multi-tests for truck passing (20 MPH, 32 km/h) [17]: (a) first test at low speed, (b) second test at low speed, (c) first test at higher speed, and (d) second test at higher speed.

This indicates that the CNT concrete sensors have stable and repeatable traffic detection capability.

10.6.2.4 Detection of Van Passing

Figures 10.42 and 10.43 give the detection results of the van passing at low and higher speeds. It can be seen from Figure 10.42(a) and (b), and Figure 10.43(a) and (b) that changes occur in the voltage signal curves of the cast-in-place and pre-cast CNT concrete sensors when the van passes over them. This indicates that the CNT concrete sensors can identify the front wheel and the rear wheel passing of the van. However, as shown in Figure 10.42(c) and (d), and Figure 10.43(c) and (d), the van passing cannot be detected by the strain gauges. This is because the van loading is much lower than the truck loading.

In the road tests, both polarization of CNT concrete under external electrical field [4] and changes in environmental temperature and humidity (i.e., the moisture content of the sensors) [21] are found to have some effect on the electrical resistivity of the CNT concrete sensors. However, the changes in electrical resistivity signals (i.e., voltage signals) caused by the polarization and environmental factors are

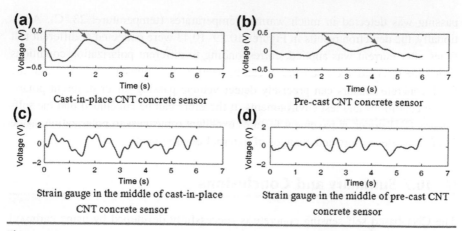

Figure 10.42 Detection results of van passing at low speed (5 MPH, 8 km/h) [17], (a) Cast-in-place CNT concrete sensor, (b) Pre-cast CNT concrete sensor, (c) Strain gauge in the middle of cast-in-place CNT concrete sensor, (d) Strain gauge in the middle of pre-cast CNT concrete sensor.

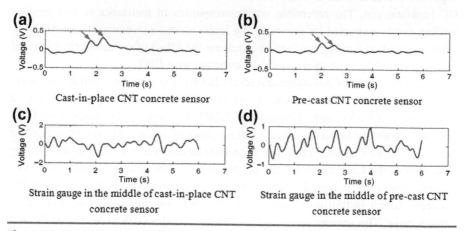

Figure 10.43 Detection results of van passing at higher speed (20 MPH, 32 km/h) [17], (a) Cast-in-place CNT concrete sensor, (b) Pre-cast CNT concrete sensor, (c) Strain gauge in the middle of cast-in-place CNT concrete sensor, (d) Strain gauge in the middle of pre-cast CNT concrete sensor.

continuous and gradual, while those caused by vehicular loading are transient and abrupt. Therefore, the former can be filtered out in the post-processing of measured voltage signals, and they will not influence the detection accuracy of the CNT concrete sensors. In addition, it should be mentioned that the detection of truck passing was performed in colder temperatures (temperature: 3.5 °C), while the van passing was detected in much warmer temperatures (temperature: 23 °C). Additionally, the detection results in Figures 10.39–10.43 were collected at different test times after current was applied, corresponding to different polarization conditions inside the CNT concrete sensors. It can be seen from Figures 10.39–10.43 that the CNT concrete sensors can precisely detect vehicle passing under different polarization conditions and test environments. It therefore can be concluded that the self-sensing CNT concrete pavement features excellent robustness to polarization inside CNT concrete sensors and changes of external environments.

10.7 Summary and Conclusions

The CNT-based self-sensing concrete is successfully developed by using carboxyl CNT as filler and NaDDBS and SDS as surfactants, which are used to improve the dispersion of CNT in concrete matrix.

The concrete with CNT has both resistance and capacitance characteristics. The response of capacitance to compressive loading is insensitive, but the charging of capacitor causes a linear increase component in the measured resistance during the DC measurement. The reversible sensing responses of resistance to compressive loading can be extracted by removing the linear increase component. An AC measurement method can also be used to eliminate the effect of capacitor charging and discharging on the sensing responses of concrete with CNT.

Concrete with CNT has stable and reversible sensing responses within the elastic regime, and can achieve a sensitivity of 0.911 kΩ/MPa, a linearity of 7.16%, a repeatability of 1.53%, and a hysteresis of 7.24%. The sensing responses of the CNT concrete are almost free from the effect of loading rate when the loading rate is lower than 0.20 cm/min. If the loading rate exceeds 0.20 cm/min, it will affect the sensing responses of the CNT concrete, and the effect increases with the loading rate. The sensing sensitivity is heavily dependent on the conductive network in the CNT concrete, which in turn is influenced by the CNT concentration level and water/cement ratio. The sensing sensitivities of concrete with 0.05 wt.%, 0.1 wt.%, and 1 wt.% of CNT first increase and then decrease with the increase of CNT concentration levels. The electrical resistance of CNT concrete with 0.6 water/cement ratio is more sensitive to compressive stress than that of concrete with 0.45 water/cement ratio. The sensing property of CNT concrete strongly depends on the water content in the concrete. The sensing sensitivities of CNT concrete with 0.1%, 1.3%, 3.3%, 5.7%, 7.6%and 9.9% water content are 0.60 kΩ/MPa, 0.61 kΩ/MPa, 0.73 kΩ/MPa, 0.68 kΩ/MPa, 0.34 kΩ/MPa, and 0.06 kΩ/MPa, respectively. Both the change in intrinsic resistance of CNT and the change in contact resistance between adjacent CNTs contribute to sensing property of the CNT concrete, but the latter is the leading contributing factor.

A CNT-based self-sensing concrete pavement system for traffic detection is integrated by embedding precast and cast-in-place self-sensing CNT concrete into pavement. This system performs very well and can be applied to a real-time online detection of vehicle passing. It has the advantages of high detection precision, high anti-jamming ability, easy installation and maintenance, long service life, and good structural property.

References

[1] http://jnm.snmjournals.org/content/48/7/1039/F1.expansion.html.

[2] http://www.azonano.com/article.aspx?ArticleID=2180.

[3] http://cnx.org/content/m22580/latest/?collection=col10700/latest.

[4] Han BG, Yu X, Ou JP. Multifunctional and smart carbon nanotube reinforced cement-based materials. In: Gopalakrishnan K, Birgisson B, Taylor P, Attoh-Okine NO, editors. Nanotechnology in civil infrastructure: a paradigm shift, 1–47. Publisher: Springer; 2011. p. 276.

[5] Li GY, Wang PM, Zhao XH. Pressure-sensitive properties and microstructure of carbon nanotube reinforced cement composites. Cem Concr Compos 2007;29:377–82.

[6] Saafi M. Wireless and embedded carbon nanotube networks for damage detection in concrete structures. Nanotechnology 2009;20:395502 (7pp).

[7] Cwirzen A, Habermehl-Cwirzen K, Penttala V. Surface decoration of carbon nanotubes and mechanical properties of cement/carbon nanotube composites. Adv Cem Res 2008;20:65–73.

[8] Makar J, Margeson J, Luh J. Carbon nanotube/cement composites-early results and potential applications. In: Third International conference on construction materials: performance. Vancouver: Innovations and Structural Implications; 2005. pp. 1–10.

[9] De Ibarra YS, Gaitero JJ, Erkizia E, Campillo I. Atomic force microscopy and nanoindentation of cement pastes with nanotube dispersions. Phys Status Solidi (A) 2006;203:1076–81.

[10] Shah SP, Konsta-Gdoutos MS, Metexa ZS. Highly-dispersed carbon nanotube-reinforced cement-based materials. Patent US 20090229494A1.

[11] Han BG, Zhang K, Yu X, Kwon E, Ou JP. Electrical characteristics and pressure-sensitive response measurements of carboxyl MWNT/cement composites. Cem Concr Compos 2012;34:794–800.

[12] Han BG, Yu X, Zhang K, Kwon E, Ou JP. Sensing properties of CNT filled cement-based stress sensors. J Civ Struct Health Monit 2011;1:17–24.

[13] Han BG, Yu X, Kwon E, Ou JP. Effects of CNT concentration level and water/cement ratio on the piezoresistivity of CNT/cement composites. J Compos Mater 2012;46(1):19–25.

[14] Han BG, Yu X, Ou JP. Effect of water content on the piezoresistivity of MWNT/cement composites. J Mater Sci 2010;45:3714–9.

[15] Sun SW, Yu X, Han BG. Sensing mechanism of self-monitoring CNTs cementitious composite. J Test Eval 2014;42(1). http://dx.doi.org/10.1520/JTE20120302.

[16] Han BG, Yu X, Kwon E. A self-sensing carbon nanotube/cement composite for traffic monitoring. Nanotechnology 2009;20:445501 (5pp).

[17] Han BG, Zhang K, Burnham T, Kwon E, Yu X. Integration and road tests of a self-sensing CNT concrete pavement system for traffic detection. Smart Mater Struct 2013;22:015020 (8pp).

[18] Raki L, Beaudoin J, Alizadeh R, Makar J, Sato T. Cement and concrete nanoscience and nanotechnology. Materials 2010;3:918–42.

[19] Chen SJ, Collins FG, Macleod AJN, Pan Z, Duan WH, Wang CM. Carbon nanotube-cement composites: a retrospect. IES J Part A: Civ Struct Eng 2011;4(4):254–65.

[20] Islam MF, Rojas E, Bergey DM, Johnson AT, Yodh AG. High weight fraction surfactant solubilization of single-wall carbon nanotubes in water. Nano Lett 2003;3(2):269–73.

[21] Vaisman L, Wagner HD, Marom G. The role of surfactants in dispersion of carbon nanotubes. Adv Colloid Interface Sci; 2006:128–30. 37–46.

[22] Azhari F. Cement-based sensors for structural health monitoring [dissertation for the Master Degree of Applied Science]. Vancouver (Canada): University of British Columbia; 2008.

[23] Hou TC, Lynch JP. Conductivity-based strain monitoring and damage characterization of fiber reinforced cementitious structural components. Proc SPIE 2005;5765:7–10.

[24] Richardson AE. Electrical properties of Portland cement, with the addition of polypropylene fibres-regarding durability. Struct Surv 1983;22(3):156–63.

[25] Sun MQ, Li ZQ, Liu QP. The electromechanical effect of carbon fiber reinforced cement. Carbon 2000;40(12):2263–4.

[26] Torrents JM, Mason TO, Peled A, Shah SP, Garboczi EJ. Analysis of the impedance spectra of short conductive fiber-reinforced composites. J Mater Sci 2001;36:4003–12.

[27] Tumidajski PJ. Electrical conductivity of Portland cement mortars. Cem Concr Res 1996;26(4):529–43.

[28] Song GL. Equivalent circuit model for AC electrochemical impendance spectroscopy of concrete. Cem Concr Res 2000;30:1723–30.

[29] Chen B, Wu KR, Yao W. Conductivity of carbon fiber reinforced cement-based composites. Cem Concr Compos 2004;26(4):291–7.

[30] Mao QZ, Zhao BY, Shen DR, Li ZQ. Influence of polarization on conductivity of carbon fiber reinforced cement. Chin J Mater Res 1997;11(2):195–8.

[31] Shi ML, Chen ZY. Study of AC impedance on the porous structure of hardened cement paste. J Build Mater 1998;1(1):30–5.

[32] Li CY, Chou TW. Modeling of damage sensing in fiber composites using carbon nanotube networks. Compos Sci Technol 2008;68:3373–9.

[33] Han BG. Properties, sensors and structures of pressure-sensitive carbon fiber cement paste [dissertation for the Doctor Degree in Engineering]. Harbin Institute of Technology; 2006.

[34] Mao QZ, Chen PH, Zhao BY, Li ZQ. Resistance changement of compression sensible cement speciment under different stresses. J Wuhan Univ Technol 1996;11(3):41–5.

[35] Chung DDL. Piezoresistive Cement-based materials for strain sensing. J Intelligent Material Syst Struct 2002;13(9):599–609.

[36] Dresselhaus MS, Dresselhaus G, Avouris PH. Carbon nanotubes: synthesis, structure, properties, and applications. New York: Springer, Verlag, Berlin, Heidelberg; 2000.

[37] Han BG, Han BZ, Ou JP. Experimental study on use of metal powder-filled cement-based composite for fabrication of piezoresistive sensors with high sensitivity. Sens Actuators Phys 2009;149:51–5.

[38] Chen B, Wu KR, Yao W. Piezoresistivity in carbon fiber reinforced cement based composites. J Mater Sci Technol 2004;20:746–50.

[39] Jiang MJ, Dang ZM, Xu HP, Yao SH. Effect of aspect ratio of multiwall carbon nonotubes on resistance-pressure sensitivity of rubber nanocomposites. Appl Phys Lett 2007;91:072907 (3pp).

[40] Zallen R. The Physics of Amorphous Solids. New York: Wiley; 1983.

[41] Ou JP, Han BG. Piezoresistive cement-based strain sensors and self-sensing concrete components. J Intell Mater Syst Struct 2009;20(3):329–36.

[42] Han BG, Ou JP. The humidity sensing property of cements with added carbon. New Carbon Mater 2008;23:382–4.

[43] Liu LT, Ye XY, Wu K, Han R, Zhou ZY, Cui TH. Humidity sensitivity of multi-walled carbon nanotube networks deposited by dielectrophoresis. Sensors 2009;9:1714–21.

[44] Romero HE, Sumanasekera GU, Kishore S, Eklund PC. Effects of adsorption of alcohol and water on the electrical transport of carbon nanotube bundles. J Phys Condens Matter 2004;16:1939–49.

[45] Tang DS, Ci LJ, Zhou WZ, Xie SS. Effect of H_2O adsorption on the electrical transport properties of double-walled carbon nanotubes. Carbon 2006;44:2155–9.

[46] Na PS, Kim H, So HM, Kong KJ, Chang H, Ryu BH, et al. Investigation of the humidity effect on the electrical properties of single-walled carbon nanotube transistors. Appl Phys Lett 2005; 87:093101 (3pp).

[47] Grujicic M, Gao G, Gersten B. Enhancement of field emission in carbon nanotubes through adsorption of polar molecules. Appl Surf Sci 2003;206:167–77.

[48] Chen CW, Lee MH, Clark SJ. Gas molecule effects on field emission properties of single-walled carbon nanotube. Diam Relat Mater 2004;13:1306–13.

[49] Qiao L, Zheng WT, Wen QB, Jiang Q. First-principles density-functional investigation of the effect of water on the field emission of carbon nanotubes. Nanotechnology 2007;18:155707 (5pp).

[50] Tashiro C, Ishida H, Shimamura S. Dependence of the electrical resistivity on evaporable water content in hardened cement pastes. J Mater Sci Lett 1987;6:1379–81.

[51] Lu JR, Chen XF, Lu W, Chen GH. The piezoresistive behaviors of polyethylene/foliated graphite nanocomposites. Eur Polym J 2006;42:1015–21.

[52] Wang LH, Ding TH, Wang P. Influence of carbon black concentration on piezoresistivity for carbon-black-filled silicone rubber composite. Carbon 2009;47:3151–7.

[53] Chen K, Xiong CX, Li LB, Zhou L, Lei Y, Dong LJ. Conductive mechanism of antistatic poly (ethylene terephthalate)/ZnOw composites. Polym Compos 2008;30:226–31.

[54] Wansom S, Kidner NJ, Woo LY, Mason TO. AC-impedance response of multi-walled carbon nanotube/cement composites. Cem Concr Compos 2006;26:509–19.

[55] Zhang BN, Zhang KH. A study on force-resistance relationship of smart concrete. J Chongqing Jiaotong Univ 2006;25(1):87–9.

[56] Pushparaj VL. Nalamasu. O Manoocher Birang M. Carbon nanotube-based load cells. Patent US2010/0050779 A1.2010.

[57] Tombler TW, Zhou C, Alexseyev L, Kong J, Dai H, Liu L, et al. Reversible electromechanical characteristics of carbon nanotubes under local-probe amanipulation. Nature 2000;405:769–72.

Chapter 11

Challenges of Self-Sensing Concrete

Self-Sensing Concrete in Smart Structures. http://dx.doi.org/10.1016/B978-0-12-800517-0.00011-3

11.1 Introduction and Synopsis

Smart (or intelligent) material technology represents an emerging research field that is finding many applications in civil infrastructures. These applications include condition/health monitoring, damage assessment, structural control, structural repair and maintenance, integrity assessment and more recently asset management, preservation, and operation of civil infrastructures. The potential benefits of this technology include improved infrastructures' reliability and longevity, enhanced structural performance and durability, improved safety against natural hazards and vibrations, and a reduction in life-cycle costs in operating and managing civil infrastructures.

Smart concrete represents the development direction of concrete from high strength and high performance to multifunctionality and intelligence. Self-sensing concrete is a kind of smart concrete that has attracted wide attention from academia and industry. This chapter will briefly summarize the state-of-the-art of smart concrete. It should be noted that the overall research and development in smart concrete is so broad that it is worthy of a separated book. This book is focused on the progress in self-sensing concrete. Based on this, challenges for development and deployment of self-sensing concrete are discussed.

11.2 Smart Concrete

As a type of smart material, smart concrete is a new generation of concrete materials that possess adaptive capabilities to external stimuli, such as loads or environment, with inherent intelligence.

The concept of smart concrete was first proposed by Japanese researchers in the late 1980s. Since then, Yanagida of the University of Tokyo developed "self-test concrete" by including glass and carbon fibers in concrete in 1992. Dry of the University of Illinois embedded calcium nitrite contained in polypropylene fibers in concrete for anticorrosion of steel reinforcement bars in 1992 [1–3]. This kind of "smartness" is generally programmed by material composition, special processing, introduction of other functional components, or by modification of the microstructure. The "smartness" may be characterized by self-sensing, self-heating, self-healing, self-cleaning, self-adjusting, etc. [2,3]. Self-sensing concrete has been detailed discussed in previous chapters of this book. This section will briefly summarize current research and development in other types of smart concretes.

11.2.1 SELF-HEALING CONCRETE

Self-healing concrete is mostly defined as the ability of concrete to repair its cracks autogenously or autonomously. It is also called self-repairing concrete. Cracks in concrete are a common phenomenon due to its relatively low tensile strength. Durability of concrete is impaired by these cracks since they provide an easy path for the transportation of liquids and gases that potentially contain harmful substances. If microcracks grow and reach the reinforcement, not only the concrete itself may be attacked, but also the reinforcement steel bars will be corroded. Therefore, it is important to control the crack width and to heal the cracks as soon as possible. Self-healing of cracks in concrete would contribute to a longer service life of concrete structures and would make the material not only more durable but also more sustainable [4–6].

Self-healing is actually an old and well-known phenomenon for concrete as it possesses some natural autogenous healing properties. Due to ongoing hydration of clinker minerals or carbonation of calcium hydroxide ($Ca(OH)_2$), cracks may heal after some time. However, autogenous healing is limited to small cracks and is only effective when water is available, thus making it difficult to control. Nonetheless, concrete may be modified to build in autonomous crack healing [4]. Dry started to work on the autonomous self-healing concrete in 1994 [5]. In the following years, several researchers started to investigate this topic. Many self-healing approaches are proposed. They mainly include autogenous self-healing method, capsule-based self-healing method, vascular self-healing method, electrodeposition self-healing method, microbial self-healing method, and self-healing method through embedding shape memory alloys (SMAs) [4–6]. For example, Edvardsen found that the greatest potential for autogenous healing exists in early age concrete [7]. Mihashi et al. used urea-formaldehyde microcapsules (diameter 20–70 μm) filled with epoxy resin and gelatin microcapsules (diameter 125–297 μm) filled with acrylic resin to achieve self-healing of concrete under compression and splitting [8]. Joseph et al. made use of an air-curing healing agent, provided by glass tubes. One end of the tubes was open to the atmosphere and curved to supply healing agent. When the tubes become depleted after concrete cracking occurred, additional agent could be added via the open end to allow healing of wider cracks [9]. Otsuki et al. proposed the electrodeposition method as a means of repair for cracked concrete structures and investigate the effects of this method on various concrete properties [10]. Jonkers et al. investigated the potential of bacteria to act as self-healing agent in concrete, i.e., their ability to repair occurring cracks. They proved that application of bacterial spores as self-healing agent appears promising [11]. Kuang and Ou, and Li et al., found that the SMA wire as reinforcing bar can make cracks close and perform the task of emergency damage repair in concrete structures. The cracks are closed due to the super elastic behavior of embedded SMAs [12,13].

11.2.2 SELF-ADJUSTING CONCRETE

Self-adjusting property of concrete means the ability to perform proper responses to loading or environmental change. It mainly includes self-damping, temperature self-adjusting, and humidity self-adjusting performances. Self-adjusting concrete mitigates hazards (whether due to accidental loading, wind, ocean waves, or earthquakes), increases the comfort of people who use the structures, and enhances the reliability and performance of structures. It has a very wide range of applications in civil engineering structures, especially for high-rise buildings, long span bridges, high-speed railways, and nuclear power stations [14–16].

In the 1990s, researchers began to try to improve the damping property of concrete by adding macromolecular material. For example, Chung and her team conducted a series of studies to improve the damping property and stiffness of concrete by using admixtures such as silica fume and latex [17,18]. Cao et al. provided a means of increasing the damping ratio of concrete by adding carboxylic styrene butadiene latex [19]. Liu and Ou observed that the addition of polymer or silane-treated silica fume can improve the loss factor of concrete, thus increasing self-damping capability of concrete [20,21]. Yao et al. pointed out that the mechanical performance and wearing resistance of concrete modified by carboxylic styrene butadiene latex are much better [22]. Luo considered low interfacial shear strength between the fibers and the matrix and high interfacial area will lead to increase in damping capacity. He studied the self-damping capacity of concrete with different concentrations of carbon nanotube (from 0.1 wt.% to 2 wt.%), and found that the critical damping ratio of nanocomposites increases with concentrations of carbon nanotube. The critical damping ratio of nanocomposites with 2% of carbon nanotube is 1.6 times than that of plain concrete [23].

Phase change materials (PCMs) have the ability to absorb or release heat when the material changes from solid to liquid and vice versa [24]. The combination of conventional concrete material and latent heat storage PCMs thus can provide temperature self-adjusting concrete, which can be used to increase thermal comfort of buildings, store thermal energy, and lower the energy consumption of buildings [24–26]. Cabeza et al. tested the performances of a small house-sized cubicle with PCM concrete and one with standard concrete. Temperature differences up to 4 °C were observed between both cubicles, and peak temperatures in the PCM concrete cubicle were shifted to later hours. The temperature self-adjusting concrete with PCM can improve thermal inertia of the building and achieve better thermal comfort inside the building [27]. Ling and Poon pointed out that concrete incorporating PCMs can provide a more stable temperature, which ensures temperatures conducive to human thermal comfort in a building's internal space can be maintained [28]. Han et al. found that a temperature difference up to 6.8 °C was observed between interiors of two same-size scaled-down building models (one made of plain concrete, the other one made of concrete with PCMs). The melting of PCMs absorbs heat energy and delays the rise of interior temperatures of the buildings, which can naturally cool down building interiors from summer heat [29].

In addition, the Japanese researchers added zeolite powder to concrete to achieve the self-adjusting of humidity in galleries or museums. The self-adjusting property of concrete results from adsorption and release of porous zeolite powder to water indoors during the process of environmental humidity change [16].

11.2.3 SELF-HEATING CONCRETE

Self-heating concrete refers to concrete material using electrical resistance heating based on the Joule effect. It is also called electrothermal concrete. Conventional concrete material is not electrically conductive. Its resistivity is too high to be effectively heated. The resistivity of concrete can be reduced by adding electrically conductive fillers such as carbon fibers, steel fibers, and graphite [30–36]. Self-heating concrete has excellent potential for domestic and outdoor environments, especially for deicing and snow-melting of parking garages, sidewalks, driveways, highway bridges, and airport runways. It would eliminate or dramatically reduce the need for using salt, thus providing an effective and environmentally friendly alternative. Self-heating concrete itself is the heating element, and thus is able to generate the heat more uniformly throughout the heated structure and has no damage to infrastructure.

Much effort has been made to study the property and application of self-heating concrete. For example, Xie et al. [31,32] first developed conductive concrete by using conductive fibers and particles as fillers. This concrete presents good conductivity and mechanical strength. When a voltage is applied to this conductive concrete, thermal energy was radiated from this concrete in the same fashion as a metallic wire conductor. Snow melting using this conductive concrete was demonstrated both in the laboratory and in the field. Yehia and Tuan [30,33–35] studied several typical concretes using steel fibers, steel shaving, and graphite as conductive fillers. They found that the conductive concrete can raise the slab temperature from −1.1 to 15.6 °C in 30 min with an average input power of approximately 520 W/m^2. Pye et al. [36] also used carbon black and coke as conductive fillers in fabricating self-heating concrete. Chuang [37] compared the self-heating efficiency of several conductive concretes. The conductive concrete with 8 μm-diameter steel fibers was found to possess lower electrical resistivity and exceptionally high heating effectiveness. The heating power per unit area attained by the steel fiber concrete was 750 W/m^2, much higher than 340 W/m^2 for a metal wire with the same resistance (note: the heating power per unit area shows the heating capability of the self-heating concrete). Zhang et al. observed that the concrete with nickel particles has a superior self-heating performance. It can achieve a temperature increment of about 50 °C within 30 s when the input voltage is 20 V. At an input voltage of 15 V, the concrete can deice 3 mm of ice in 478 s under an ambient temperature of -16.0 °C and melt 2 cm of snow in 368 s under an ambient temperature of -5.3 °C, respectively [38].

Beside these main types of smart concrete, researchers also fabricated some new types of smart concrete such as self-cleaning photocatalytic concrete [39], light-transmitting concrete [40], electromagnetic shielding/absorbing concrete [41], and energy self-harvesting concrete [42].

11.3 Stress/Strain Sensing for Concrete

The previous chapters of this book have detailed discussed self-sensing concrete that can detect its own structural health parameters. It should be noted that there is also quite a bit of research on the sensing of concrete parameters using external sensors or devices. This section will summarize those sensing technologies for concrete.

11.3.1 CONCRETE INTEGRATED WITH ELECTRIC RESISTANCE STRAIN GAUGES

Electric resistance strain gauges (also called strain gauges) are sensors whose electrical resistance varies with applied force or deformation. The strain can be measured by detecting changes in electrical resistance. Commonly, electric resistance strain gauges are attached on the concrete surface, are embedded within the concrete, or are installed on reinforcement bars within the concrete structure for strain monitoring [43]. The electric resistance strain gauges are the most mature and widely used of the electrical measurement techniques for concrete structures because they are easy to install and low cost and show excellent reproducibility of sensing property. However, the conventional resistance strain gauges exhibit a low sensitivity (i.e., gauge factor) ranging from 2 to 5, and are easily affected by electromagnetic interference. In addition, the resistance strain gauges were prone to drift, making them unsuitable for long-term monitoring [43,44].

11.3.2 CONCRETE INTEGRATED WITH OPTIC FIBERS

Optic fibers can be used as sensors to measure strain, temperature, pressure, and other quantities by modifying a fiber so that the property to measure modulates the intensity, phase, polarization, wavelength, or transit time of light in the fiber. Sensors that vary the intensity of light are the simplest, since only a simple source and detector are required. A particularly useful feature of such fiber optic sensors is that they can provide distributed sensing over distances of up to 1 m [45].

Optic fibers can be incorporated into concrete to measure strain, displacement, moisture, corrosion, crack, and temperature [46–48]. Lau et al. embedded optic fiber Bragg grating (FBG) sensors into concrete structures. The use of the embedded FBG sensor can measure strain accurately in different locations and provide information to the operator that the structure is subjected to debond or microcrack failure [49]. Lee et al. used optic fiber sensors for the measurement of crack-tip opening displacements of concrete structures [50]. Yeo et al. embedded optic fiber based humidity sensors in concrete and used them for monitoring moisture changes in concrete [51]. Fuhr and Huston presented optic fiber corrosion sensors that could be embedded in roadway and bridge structures. The degradation of reinforced concrete due to corrosion can be detected by these sensors [48]. Childs et al. embedded FBG sensors in concrete cylinders to monitor cracking depth [52]. Zhou et al. embedded optic fiber sensors in a concrete structure for sensing temperature change and concrete hydration process during the early age of hydration [53]. Glišić and Inaudi embedded optic fiber sensors in fresh concrete to monitor early and very early age deformation of concrete [54]. Kuang et al. employed plastic optic fibers to detect initial cracks, monitor post-crack vertical deflection, and detect failure cracks in concrete beams subjected to flexural loading [55]. Bernini et al. attached single-mode optic fiber sensors to the concrete beam to detect both tensile and compressive strains. The sensors are also able to detect the formation of a crack in the midsection of the concrete beam [56]. Zhou et al. integrated optic fiber sensors in fiber-reinforced polymer and embedded them in concrete pavement to monitor the 3D strain distribution of concrete [57].

Optic fibers exhibit several advantages such as high sensitivity, flexibility, embeddability, multiplexity, and electrical or magnetic interference immunity. However, optic fibers need to be carefully handled and protected to prevent damage as they are incorporated into concrete structures. Their durability is not sufficient for long-term monitoring due to optic fiber aging.

11.3.3 Concrete Integrated with Piezoelectric Materials

Piezoelectric materials can be classified in three categories: piezoelectric ceramics (PZT), piezoelectric polymers, and piezoelectric composites [58–60]. Piezoelectric materials exhibit the sensing ability resulting from piezoelectric effect i.e., a surface charge is generated in response to an applied mechanical stress (direct effect) and conversely, a mechanical strain is produced in response to an applied electric field (converse effect) [61]. The piezoelectric materials are very sensitive in detecting stress, temperature, and cracks of concrete. Wen et al. successfully monitor the temperature and stress of concrete under static or quasi-static states by embedding piezoelectric materials into concrete [62]. Soh et al. surface-bonded PZT patches to carry out the crack and damage detection during the destructive load testing of a prototype reinforced concrete bridge [63]. Saafi and Sayyah attached an array of PZT sensors at a concrete structure to detect and localize disbonds and delaminations of reinforcement from the concrete structure [64]. Song et al. detected internal cracks of a 6.1-m-long reinforced concrete bridge bent-cap by imbedding PZT inside one end of the concrete [65,66]. Zhao et al. used several PZT built in the concrete beam to monitor cracks [67]. Xu prepared 2–2 and 1–3 concrete-based PZT piezoelectric composite sensors by cutting-casting method to monitor the temperature, stress, and cracks of concrete structures adhered and embedded with these sensors [68]. Yokoyama and Harada used the piezoelectric polymer (PVDF) film to monitor the initiation and propagation of cracks and detect damage of reinforced concrete beams [69]. Wang and Yi used the PVDF-based stress gauge for stress measurement of concrete under impact [70].

The piezoelectric materials possess advantages of high sensitivity, high resonance frequency, high stability, etc. The piezoelectric materials can produce only electrical response to the dynamic mechanics. One disadvantage of piezoelectric materials is that they cannot be used for truly static measurements. In addition, the piezoelectric materials also show disadvantages in the unfavorable compatibility and poor durability with the concrete structures [71].

11.3.4 CONCRETE INTEGRATED WITH SMA

SMAs can work as sensors because their electrical resistance is dependent on their strain (the electric resistance is increased with applied tension strain) [72]. SMA can be used to monitor the strain (or deformation) of concrete and to estimate the crack width in concrete [73,74]. For example, Song et al. conducted bending tests on concrete beams reinforced with SMA cables. It was found that the electrical resistance value of the SMA cable experienced large and repeatable changes with the opening and closing of the crack, indicating that electrical resistance can be used to monitor crack width [73]. Liu studied the sensing property of SMA wires. The tested results show that the fracture in electric resistance changes almost linearly with the strain, and the relationship between fracture of electric resistance and stress is similar to the relationship between the stress and strain. He also investigated the sensing behavior of concrete beams reinforced with SMA wires. The rate of change of SMA resistance has a good linear relationship with the mid-span deflection of the beam. The embedded SMA wire will monitor strain and deformation of concrete structures [74]. The SMA presents good durability and corrosion resistance, while it has low sensitivity (i.e., gauge factor) ranging from 3.8 to 6.2 and high cost [74].

11.3.5 CONCRETE INTEGRATED WITH SELF-DIAGNOSING FIBER-REINFORCED POLYMER COMPOSITES

Self-diagnosing (or self-monitoring) fiber-reinforced composites contain an electrical conductive phase such as carbon fiber (short carbon fiber or continuous carbon fiber) and conductive powder (e.g., graphite powder, carbon black) in the polymer matrix. The external force can lead to regular change in electrical resistivity of self-diagnosing fiber-reinforced composites, so they have the abilities to monitor strain (or deformation), crack, and damage of concrete by measuring electrical resistance [75–78]. The polymer composites are then embedded in concrete structures as sensing elements. The sensing mechanism of these polymer composites is similar to the self-sensing concrete discussed in this book. However, the polymer matrix based composite still works as embedded sensors while the self-sensing concrete itself works as the sensor. For example, Hiroshi et al. fabricated self-diagnosing fiber-reinforced composites with the ability to memorize damage history, and bonded composite film sensors on the concrete surface to detect cracks and measure crack width of the reinforced concrete bridge pier columns under quasistatic cyclic lateral loading. They successfully detected damage to concrete structures through confirmation of the relationship between the extent of damage and the variation of electrical conductivity of self-diagnosis materials [75]. Sugita et al. embedded carbon fiber and glass fiber reinforced polymer materials in concrete structures as reinforcement and sensing materials. They observed that the electrical resistance characteristics of reinforced concrete change along with changing loads. A permanent residual electrical resistance could be observed after the removal of load, and its change was dependent on the maximum load applied. Monitoring changes in electrical resistance during and after loading is thus a promising method for anticipating the fracture of the reinforced concrete [76]. Muto et al. stated that the self-diagnosis carbon fiber and glass fiber reinforced polymer materials can be used to give early warnings of catastrophic failure of concrete structures, and monitor high values of strain. They also embedded self-diagnosis polymer material grid in the 20th floor slab of a skyscraper to detect the occurrence of cracks in concrete [77]. Yang et al. fabricated self-diagnosis hybrid carbon fiber reinforced polymer sensors. These sensors can monitor the whole loading procedure of concrete structures with high sensitivity, including the elastic deformation, the yielding of reinforcing steel bars, and the initiation and propagation of cracks in concrete [78].

The most obvious advantage of self-diagnosing fiber reinforced composites is that they work as both structural materials and sensing materials. However, the sensing repeatability and stability of these composites are heavily affected by aging of polymer matrix. In addition, the self-diagnosing fiber-reinforced composites exhibit a relatively low sensitivity ranging from 30 to 40 [79].

To sum up, the external sensors or sensing materials have such drawbacks as poor durability, low sensitivity, high cost, low survival rate, unfavorable compatibility with concrete, and the need for expensive peripheral equipment, such as electronics and lasers. These mainly lead to insufficient service life and weakening of mechanical property and durability of concrete structures.

The intrinsic self-sensing concrete that has been discussed in previous chapters of this book has both structural and sensing functions since it is made of concrete material, so it replaces the need for embedded or attached external sensors or sensing materials. The intrinsic self-sensing concrete possesses many advantages, including high sensitivity, good mechanical property, long survival time, and easy installation and maintenance compared to external sensors or sensing materials for concrete.

11.4 Challenges for Development and Deployment of Self-Sensing Concrete

Although the intrinsic self-sensing concrete emerged nearly two decades ago, the research on it still continues. Many efforts are needed to promote the development and deployment of self-sensing concrete. The following challenges are believed to be critical in the future development and deployment of this kind of materials.

11.4.1 FABRICATION OF SELF-SENSING CONCRETE

Considering the cost of functional fillers, the amount of functional fillers used in the self-sensing concrete should be as low as possible to decrease the product cost. Therefore, some useful, but not complicated technologies should be developed to fabricate the intrinsic self-sensing concrete with low concentration of functional fillers. A key solution to this issue is to find some simple, repeatable, large-scale, and low-energy consumption methods for distributing functional fillers in concrete without altering the normal manufacturing process of concrete materials. The concrete with aggregates is much more useful in real applications than that without aggregate (i.e., cement paste). But aggregates, especially coarse aggregates, will make it more difficult and complex to enhance the sensing properties of the composites by using a low concentration of functional fillers. For this challenge, the hybrid filler multi-scale composite technology and use of conductive aggregates may be the potential solutions. In addition, it is necessary to develop some simple and convenient evaluation methods for dispersion quality of functional fillers in concrete matrix. In general, it is suggested that the subsequent researches should be aimed at establishing a uniform method, guidance, and specification for design, optimization, and fabrication of the self-sensing concrete.

11.4.2 MEASUREMENT OF SENSING SIGNAL OF SELF-SENSING CONCRETE

Sensing signals contain the information reflecting conditions of the self-sensing concrete. However, measurement noise and environment uncertainty might drown effective sensing signals if no suitable sensing signal measurement and processing method is used. Therefore, it is necessary to develop some new electrode design methods, measurement circuit design methods, signal acquisition, and processing methods for wholly, deeply, and accurately mining the effective information reflecting structural conditions of the self-sensing concrete. The ultimate objective is to develop standard measurement methods and measurement equipment for the self-sensing concrete.

11.4.3 SENSING PROPERTY AND ITS MECHANISM OF SELF-SENSING CONCRETE

Previous research efforts mainly concentrate on the sensing property of the self-sensing concrete under uniaxial compressive loading. In real structures, the self-sensing concrete would be subject to different styles of loading and complex stress conditions. Therefore, the overall sensing property of the self-sensing concrete should be determined for structural applications. In future work, the "finger characteristics" and sensing performance parameters (e.g., input/output range, linearity, repeatability, hysteresis, signal to noise ratio, and zero shifts) on the one-, two-, and three-axis direction of the self-sensing concrete under complex stress condition should be obtained. Moreover, the long-term evolution law of sensing property of the self-sensing concrete should be explored. Although researchers have given much reasonable explanation on sensing property of the self-sensing concrete through combinations of experiments and theories, some explanation is still unclear or needs to be further verified because the actual mechanism is quite complex in nature. The use of some advanced analyzing and testing instruments is helpful for investigating the mechanism of sensing property under different temporal and spatial conditions. In addition, in previous research, the pressure-sensitive characteristic models are built only for the self-sensing concrete under uniaxial compression or tension. In future work, much effort should be invested into developing universal models for describing and forecasting the sensing properties of the self-sensing concrete in different temporal and spatial conditions based on experiments and numerical simulation.

11.4.4 FUNCTION EXPANSION OF SELF-SENSING CONCRETE

The existing self-sensing concrete can detect/monitor some mechanical property of concrete, including strain, stress, crack, and damage. The self-sensing concrete should have the ability of detecting/monitoring other properties of concrete, such as hydration processes, temperature, humidity, and durability. In addition, as a kind of smart materials, it is desirable that intrinsic self-sensing concrete has other intelligent or functional features, such as self-healing, self-adjusting, electromagnetic shielding/absorbing, and energy self-harvesting. This endows self-sensing concrete with high level of intelligence and multifunctionality. For this goal, new composite technology, nanotechnology, and bionic technology may provide the potential solutions.

11.4.5 APPLICATION OF SELF-SENSING CONCRETE

The self-sensing concrete not only can be used in traffic detection, but also has potential in structural health monitoring of civil infrastructures such as high-rise buildings, large-span bridges, tunnel, high-speed railways, offshore structures, dams, and nuclear power plants. Moreover, it also can be used for border or military security. There should be some profound investigations on application of the self-sensing concrete in the above-mentioned fields. Some novel retrofits of the intrinsic self-sensing concrete in structural application should be developed. In general, it is suggested that the subsequent studies on application of the intrinsic self-sensing concrete should be aimed at establishing a uniform method, guidance, and specification for design and construction of self-sensing concrete structures.

11.5 Summary and Conclusions

Smart concrete is a very broad category of material that includes self-sensing concrete, self-adjusting concrete, self-healing concrete, etc. Self-sensing concrete is a branch of smart concrete, which was the earliest proposed and has been systematically and deeply investigated. self-sensing concrete mainly has excellent mechanical property and durability, long service life, and easy installation and maintenance. Self-sensing concrete has a wide application in civil infrastructures such as high-rise buildings, highway, bridges, runways for airport, continuous slab-type sleepers for high-speed trains, dams, and nuclear power plants, and especially has great potential in the field of structural health monitoring, traffic detection, and border/military security. This would be helpful for ensuring structural integrity and safety, extending the life span of structures, improving the traffic safety and efficiency, guiding the structural and traffic design, decreasing the resource and energy consumption, etc. Self-sensing concrete is a "smart" choice for maintaining sustainable development in concrete materials and structures. It will bring a deep revolution to the field of conventional concrete materials, which should have a beneficial impact on economics, society, and environment.

References

[1] Chen PW, Chung DDL. Carbon fiber reinforced concrete as a smart material capable of non-destructive flaw detection. Smart Mater Struct 1993;2:22–30.

[2] Kamila S. Introduciton, classification and applications of smart materials: an overview. Am J Appl Sci 2013;10(8):876–80.

[3] Han BG. Properties, sensors and structures of pressure-sensitive carbon fiber cement paste. Dissertation for the Doctor Degree in Engineering. China: Harbin Institute of Technology; 2005.

[4] Abrams D. Autogeneous healing of concrete. Concrete 1925;27(2):50.

[5] Dry C. Matrix cracking repair and filling using active and passive modes for smart timed release of chemicals from fibers into cement matrices. Smart Mater Struct 1994;3(2):118–23.

[6] Tittelboom KV, Belie ND. Self-healing in cementitious materials-a review. Materials 2013;6:2182–217.

[7] Edvardsen C. Water permeability and autogenous healing of cracks in concrete. ACI Mater J 1999;96(4):448–54.

[8] Mihashi H, Kaneko Y, Nishiwaki T, Otsuka K. Fundamental study on development of intelligent concrete characterized by self-healing capability for strength. Trans Jpn Concr Inst 2000;22:441–50.

[9] Joseph C, Gardner DR, Jefferson AD, Lark RJ, Isaacs B. Self healing cementitious materials: a review of recent work. In: Proceedings of the institution of civil engineers construction materials, vol. 164; 2010. pp. 29–41.

[10] Otsuki N, Hisada M, Ryu JS, Banshoya E. Rehabilitation of concrete cracks by electrodeposition. Concr Int 1999;21(3):58–63.

[11] Jonkers HM, Thijssen A, Muyzer G, Copuroglu O, Schlangen E. Application of bacteria as self-healing agent for the development of sustainable concrete. Ecol Eng 2010;36(2):230–5.

[12] Kuang YC, Ou JP. Self-repairing performance of concrete beams strengthened using superelastic SMA wires in combination with adhesives released from hollow fibers. Smart Mater Struct 2008;17:1–7.

[13] Li H, Liu ZQ, Ou JP. Behavior of a simple concrete beam driven by shape memory alloy wires. Smart Mater Struct 2006;15:1039–46.

[14] Amick H, Monteiro PJM. Modification of concrete damping properties for vibration control in technology facilities. In: Proceedings of the SPIE-the international society for optical engineering, vol. 5933; 2005. pp. 1–12.

[15] http://www.energiforumdanmark.dk/fileadmin/pr_sentationer/BASF_-_Phase_Change_Material_-_Micronal_PCM_pdf.pdf.

[16] Yao W, Wu KR. Present situation and developing trend for intelligent concrete research. New Build Mater 2000;10:22–4.

[17] Fu X, Chung DDL. Vibration damping admixtures for cement. Cem Concr Res 1996;26(1):69–71.

[18] Li X, Chung DDL. Improving silica fume for concrete by surface treatment. Cem Concr Res 1998;28(4):493–8.

[19] Cao H, Chen XH, Hua JM, Hu ZM. Tests of polymer concrete used for structural vibration mitigation. J Vib Shock 2001;5:188–91.

[20] Liu TJ, Ou JP. Study on polymer cement mortar damping property. New Build Mater; 2003:7–9 (4).

[21] Liu TJ, Ou JP, Li JL. Effects of silane-treated silica fume on damping property of cement mortar. J Chin Ceram Soc 2003;31(11):1125–9.

[22] Yao HY, Liang NX, Sun LJ, Meng JY. Property study of SD622S polymer modified concrete and the analysis of modification mechanism. J Build Mater 2005;8(1):30–6.

[23] Luo JL. Fabrication and functional properties of multi-walled carbon nanotube/cement composites. Dissertation for the Doctoral Degree in Engineering. Harbin, China: Harbin Institute of Technology; 2009.

[24] Pérez-Lombard L, Ortiz J, Pout C. A review on buildings energy consumption information. Energy Build 2008;40(3):394–8.

[25] Khudhair AM, Farid MM. A review on energy conservation in building applications with thermal storage by latent heat using phase change materials. Energy Convers Manag 2004;45(2):263–75.

[26] Bentz D, Turpin R. Potential applications of phase change materials in concrete technology. Cem Concr Compos 2007;29(7):527–32.

[27] Cabeza LF, Castellon C, Bogues M, Medrano M, Leppers R, Zubillage O. Use of microencapsulated PCM in concrete walls for energy savings. Energy Build 2007;39:113–9.

[28] Ling TC, Poon CS. Use of phase change materials for thermal heat storage in concrete-an review. Constr Build Mater 2013;46:55–62.

[29] Han BG, Zhang K, Yu X. Enhance the thermal storage of cement-based composites with phase change materials and carbon nanotubes. J Sol Energy Eng–Trans ASME 2013;135(2):1–5. 024505.

[30] Yehia S, Tuan CY. Conductive concrete overlay for bridge deck deicing. ACI Mater J 1999;96(3):382–91.

[31] Xie P, Gu P, Beaudion JJ. Conductive concrete cement-based compositions. U.S. Patent 5,447,564. 1995;10 pp.

[32] Xie P, Beaudion JJ. Electrically conductive concrete and its application in deicing. Advances in concrete technology. In: Proceedings, second CANMET/ACI international symposium, SP-154. Farmington Hills, Mich.: American Concrete Institute; 1995. pp. 399–417.

[33] Yehia S, Tuan CY, Ferdonetal D. Conductive concrete overlay for bridge deck deicing: mixture proportioning optimization and properties. ACI Mater J 2000;97(2):172–81.

[34] Tuan CY. Electrical resistance heating of conductive concrete containing steel fibers and shavings. ACI Mater J 2004;101(1):65–71.

[35] Tuan CY, Yehia S. Evaluation of electrically conductive concrete containing carbon products for deicing. ACI Mater J 2004;101(4):287–93.

[36] Pye GB, Myers RE, Arnott MR. Conductive concrete compositions containing carbonaceous particles. Chem Abstr 2001;58:76.

[37] Chuang DDL. Self-heating structural materials. Smart Mater Struct 2004;13:562–5.

[38] Zhang K, Han BG, Yu X. Nickel particle based electrical resistance heating cementitious composites. Cold Reg Sci Technol 2011;69(1):64–9.

[39] Beeldens A. An environmental friendly solution for air purification and self-cleaning effect: the application of TiO_2 as photocatalyst in concrete. In: Proceedings of transport research arena Europe–TRA, Göteborg, Sweden; June 2006.

[40] Kashiyani BK, Raina V, Pitroda J, Shah BK. A study on transparent concrete: a novel architectural material to explore construction sector. Int J Eng Innovative Technol 2013;2(8):83–6.

[41] Ou JP, Gao XS, Han BG. Absorption property and stealthy effectiveness of carbon fiber cement based materials. J Chin Ceram Soc 2006;34(8):901–7.

[42] http://www.israelexporter.com/Innowattech.

[43] http://www.vishaypg.com/docs/11091/tt611.pdf.

[44] Nawy EG. Concrete construction engineering handbook. Boca Raton (NY): CRS Press; 1997.

[45] http://en.wikipedia.org/wiki/Optical_fiber.

[46] Hollaway LC. A review of the present and future utilisation of FRP composites in the civil infrastructure with reference to their important in-service properties. Constr Build Mater 2010;24(12):2419–45.

[47] Kuang KSC, Quek ST, Koh CG, Cantwell WJ, Scully PJ. Plastic optical fibre sensors for structural health monitoring: a review of recent progress. J Sensors; 2009:1–13.

[48] Fuhr PL, Huston DR. Corrosion detection in reinforced concrete roadways and bridges via embedded fiber optic sensors. Smart Mater Struct 1998;7(2):217–28.

[49] Lau K, Chan C, Zhou L, Wei J. Strain monitoring in composite-strengthened concrete structures using optical fibre sensors. Compos Part B: Eng 2001;32(1):33–45.

[50] Lee I, Yuan L, Ansari F, Ding H. Fiber-optic crack-tip opening displacement sensor for concrete. Cem Concr Compos 1997;19:59–68.

[51] Yeo TL, Eckstein D, McKinley B, Boswell LF, Sun T, Grattan KTV. Demonstration of a fibre-optic sensing technique for the measurement of moisture absorption in concrete. Smart Mater Struct 2006;15(2):N40.

[52] Childs P, Wong ACL, Terry W, Peng GD. Measurement of crack formation in concrete using embedded optical fibre sensors and differential strain analysis. Meas Sci Technol 2008;19(6):065301.

[53] Zou XT, Chao A, Tian Y, Wu N, Zhang HT, Yu TY, et al. An experimental study on the concrete hydration process using fabry-perot fiber optic temperature sensors. Measurement 2012;45(5):1077–82.

[54] Glišić B, Inaudi D. Monitoring of early and very early age deformation of concrete using fiber optic sensors. In: Proceedings of the second fib international congress, Naples, Italy; June 2006. p. 12. ID17–26.

[55] Kuang KSC, Akmaludding, Cantwell WJ, Thomas C. Crack detection and vertical deflection monitoring in concrete beams using plastic optical fibre sensors. Meas Sci Technol 2003;14:205–16.

[56] Bernini R, Minardo A, Ciaramella S, Minutolo V, Zeni L. Distributed strain measurement along a concrete beam via stimulated brillouin scattering in optical fibers. Int J Geophys; 2011:5. 2011, ID 710941.

[57] Zhou Z, Liu WQ, Huang Y, Wang HP, He JP, Huang MH, et al. Optical fiber Bragg grating sensor assembly for 3D strain monitoring and its case study in highway pavement. Mech Syst Signal Process 2012;28:36–49.

[58] Zheng Z, Qu Y, Ma W, Hou F. Electrical properties and applications of ceramic-polymer composites. Acta Mater Compos Sin 1998;15(4):14–9.

[59] Furukawa T, Ishida K, Fukada E. Piezoelectric properties in thee composites systems of polymer and PZT ceramics. J Appl Phys 1979;50(7):4904–12.

[60] Safari A, Sa-Gong G, Giniewicz J, Newnham RE. Composite piezoelectric sensors. In: Tressler RE, Messing GL, Pantano CG, Newnham RE, editors. Tailoring multiphase and composite ceramics. US: Springer; 1986. pp. 445–54.

[61] Shin SW, Qureshi AR, Lee JY, Yun CB. Piezoelectric sensor based nondestructive active monitoring of strength gain in concrete. Smart Mater Struct 2008;17(5):055002.

[62] Wen Y, Chen Y, Li P, Jiang DY, Guo H. Smart concrete with embedded piezoelectric devices: Implementation and characterization. J Intell Mater Syst Struct 2007;18(3):265–74.

[63] Soh CK, Tseng KKH, Bhalla S, Gupta A. Performance of smart piezoceramic patches in health monitoring of a RC bridge. Smart Materials Struct 2000;9(4):533–42.

[64] Saafi M, Sayyah T. Health monitoring of concrete structures strengthened with advanced composite materials using piezoelectric transducers. Compos Part B: Eng 2001;32(4):333–42.

[65] Song G, Mo YL, Otero K, Hu H. Develop intelligent reinforced concrete structures using shape memory alloys and piezoceramics. In: Proceedings of the third international conference on earthquake engineering. Nanjing; 2004. pp. 851–6.

[66] Song G, Gu H, Mo YL, Hsu T, Dhonde H, Zhu RRH. Health monitoring of a concrete structure using piezoceramic materials. Smart Struct Mater Int Soc Opt Photonics; 2005:108–19.

[67] Zhao X, Li H, Du D, Wang J. Concrete structure monitoring based on built-in piezoelectric ceramic transducers. In: The 15th International Symposium on: smart structures and materials & nondestructive evaluation and health monitoring. International society for optics and photonics; 2008. pp. 6932081–8.

[68] Xu DY. Fabrication and properties of cement based piezoelectric sensor and its application research in civil engineering fields. Doctoral Dissertation. Shandong University; 2010.

[69] Yokoyama K, Harada T. Structural monitoring using piezoelectric film. In: Proceedings of the 24th US-Japan bridge engineering workshop, Ninneapolis, Minnesota, USA; 2008. pp. 245–53.

[70] Meng Y, Yi WJ. Application of a PVDF-based stress gauge in determining dynamic stress-strain curves of concrete under impact testing. Smart Mater Struct 2011;20(6):065004.

[71] Huang SF, Ye ZM, Wang SH, Xu DY, Chang J, Cheng X. Fabrication and properties of 1-3 cement based piezoelectric composites. Acta Mater Compos Sin 2007;24(1):122–6.

[72] Funakubo H. Shape memory alloysIn Precision Machinery and Robotics, vol. 1. New York: Gordon and Breach; 1987.

[73] Song G, Mo YL, Otero K, Gu H. Health monitoring and rehabilitation of a concrete structure using intelligent materials. Smart Mater Struct 2006;15(2):309–14.

[74] Liu ZQ. Study on damage self-monitoring and self-repair of SMA smart concrete beam. Dissertation for the Doctor Degree in Engineering. China: Harbin Institute of Technology; 2006.

[75] Hiroshi I, Yoshiki O, Hitoshi K. Experimental study on structural health monitoring of RC columns using self-diagnosis materials. Proc SPIE 2004;5391:609–17.

[76] Sugita M, Yanagida H, Muto N. Materials design for self-diagnosis of fracture in CFGFRP composite reinforcement. Smart Mater Structrues 1995;4:A52–7.

[77] Muto N, Arai Y, Shin SG, Matsubara H, Yanagida H, Sugita M, et al. Hybrid composites with self-diagnosing function for preventing fatal fracture. Compos Sci Technol 2001;61(6):875–83.

[78] Yang C, Wu Z, Zhang Y. Structural health monitoring of an existing PC box girder bridge with distributed HCFRP sensors in a destructive test. Smart Mater Struct 2008;17(3):1–10. 035032.

[79] Wang B, Ou JP, Zhang XY, He Z. Experimental research on sensing properties of CFRP bar and concrete beams reinforced with CFRP bars. J Harbin Inst Technol 2007;39(2):220–4.

Index

Note: Page numbers followed by f indicate figures; t, tables; b, boxes.

377

Printed and bound by CPI Group (UK) Ltd, Croydon, CR0 4YY

03/10/2024

01040427-0008